Solidworks Simulation for Real Weldments

Copyright 2019,

Stone Lake Analytics, LLC

Seneca, South Carolina

www.stonelakeanalytics.com

ISBN 978-0-9963101-7-8 (paperback)

ISBN 978-1-7335955-0-6 (e-book)

Solidworks Simulation for Real Weldments

Contents

"Because That's Where the Money Is"

It is probable that Willie Sutton never uttered the apocryphal response to the question, "Why do you rob banks?" It is certain that if CAE software makers asked engineers, "Why do you want to analyze the welds?" the answer would be, "Because that's where the stress is!"

Design tools, including Solidworks, are brimming with features to help manage cut lists, joints, weld schedules, and such to help real weldments get built. We are left wanting for an easy way to put true representations of welded structures into our simulation tools. Peak stresses in many weldments can be expected at welded joints, and the diligent analyst wants to provide good answers about their structural response.

We don't think there is an easy way, but we do have a plan. Common solid operations in Solidworks can be used to model welds explicitly. The model needs to reflect both where the weld bead adds material, how it penetrates into the joint, and where mating parts are *not* joined. Modern computers and solvers can quickly generate good linear static results on such detailed models.

Strain gage on cracked weld bead; simulation had driven gage placement

Scope

Considered in this book are single metallic weldments simulated in linear static analysis.

All parts are treated as three dimensional solids – this is key for stress prediction at and near the welds. Mixed element types can be used in a large complicated weldment, but the detail we are looking for comes from true modeling of the welded joints – in stubborn detail.

The methods shown here can be used within larger articulated assemblies, with moving parts, connectors, and contact conditions. Selective modeling of weld detail where it's most needed, and disciplined use of mesh refinement, can put focus on the most important details of a loaded structure.

Linear Static Analysis

Welding almost always involves joining of different materials. Even if the base parts are of identical alloy, the filler will be a specialized welding material. For most of these examples here we will be treating all solids as one uniform material.

It is important to remember that linear static stress analysis takes only two properties as input:
- Elastic modulus (Young's modulus, E)
- Poisson's ratio (v)

These are taken to be constants. Strength properties are not used in calculation of strains and stresses. They are used only in evaluating the results.

For most families of metals, Young's modulus and Poisson's ratio are nearly the same across the family. Wrought low-alloy steels are often joined to wrought low-alloy steels, with weld filler alloyed to be forgiving over a range of cooling rates. The relevant elastic properties of these materials are usually not meaningfully different.

The strengths of these materials may be very different, e.g. 70 ksi weld wire used to join A36 steel plate. When we model the welds accurately, the analyst will have the ability to point at the weld bead surface independently of the base parts.

In some cases parts with different elastic properties are joined, such as iron castings (v ~.25) to plate steel (v ~.30). A later section of this book will document use of solid body splits and bonded contact to simulate this if it is thought necessary.

Anatomy of a Welded Joint

Sometimes parts to be welded are designed with special preparations so that weld material completely fills the space between the parts, making a continuous solid. More often simple flat ends are butted together, and weld bead is built up around the outside of the contact surface. In these cases, some or most of the contact surface will remain un-joined, or gapped.

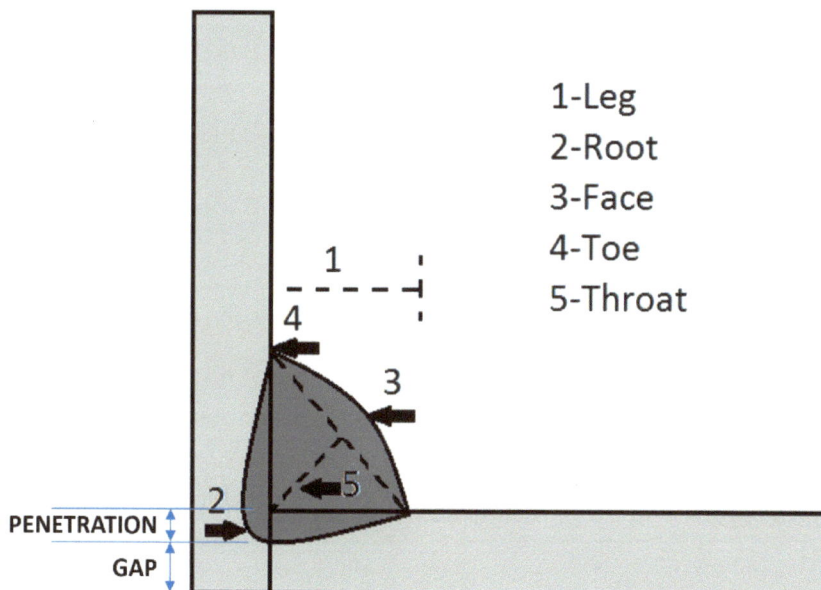

1-Leg
2-Root
3-Face
4-Toe
5-Throat

In this simplest example, a fillet weld is made on one side of a flat plate which meets another plate at a rectangular edge. The weld process melts the base material to some depth, or penetration, mixing and joining the parts into essentially one solid mass.

When we try to simulate welds accurately, we will look to replicate the gap and the penetration in our models.

Studies

Study 1: Tube End

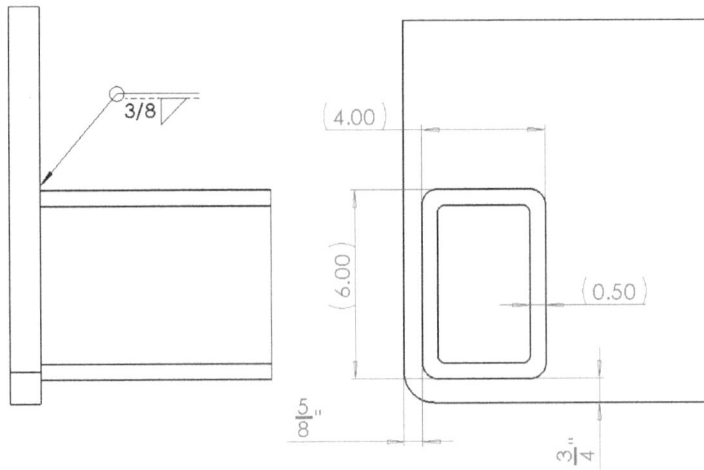

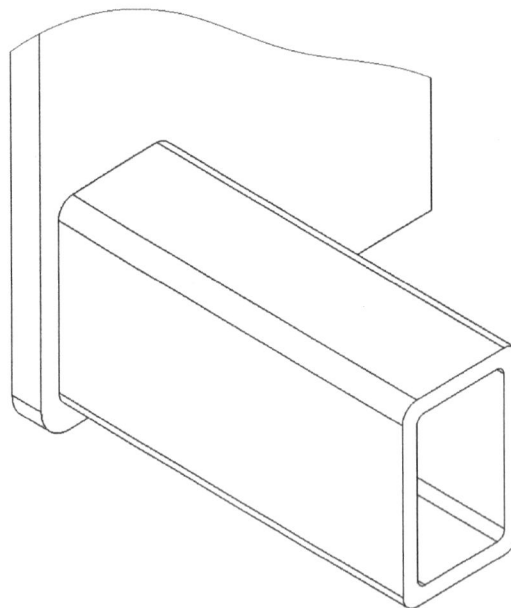

Drawing of weldment

We start with a portion of a drawing where two parts are joined with a single fillet weld, in a closed loop.

The model here has been made in a single Solidworks part. The same solid could be obtained by "join"-ing individual parts in an assembly.

The resulting butt joint is defined by two rectangular loops, inside and outside the tube where it meets the end plate.

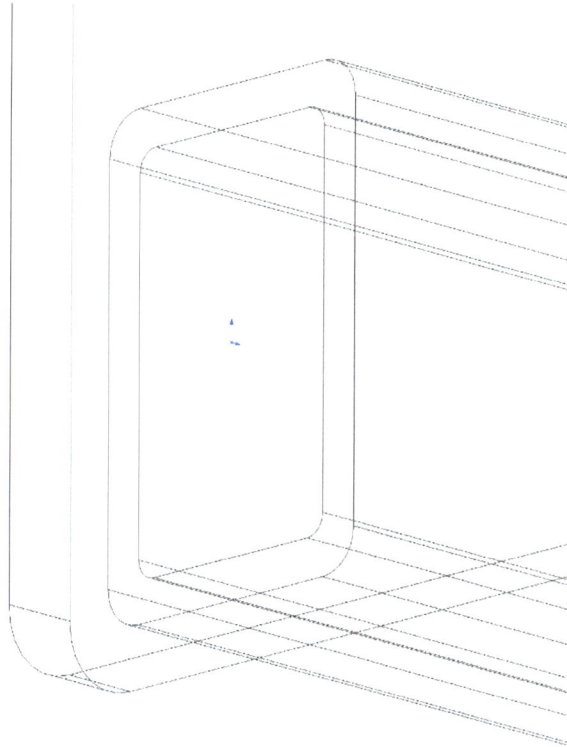

Edge view of joined parts

In simulation, this geometry is identical to the result if one ran the assembly with bonded contact, treating the mating surface as a union. So we can set up the first simulation with this shape directly.

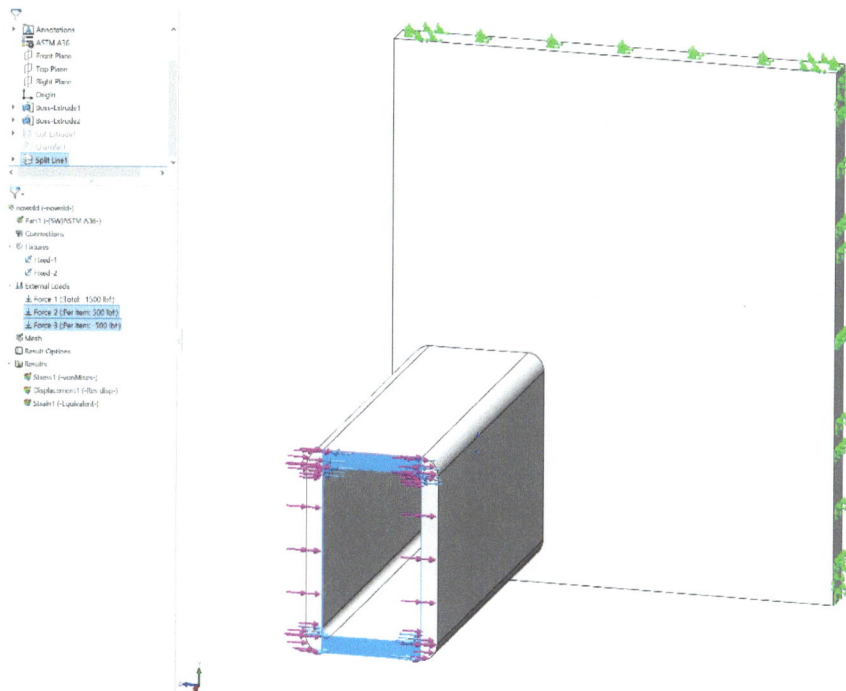

Simulation setup

To test the response of the joint a shear force (1500 lb) is applied to the end of the tube, along with a simple moment (2750 ft-lb). The edges of the plate are fixed.

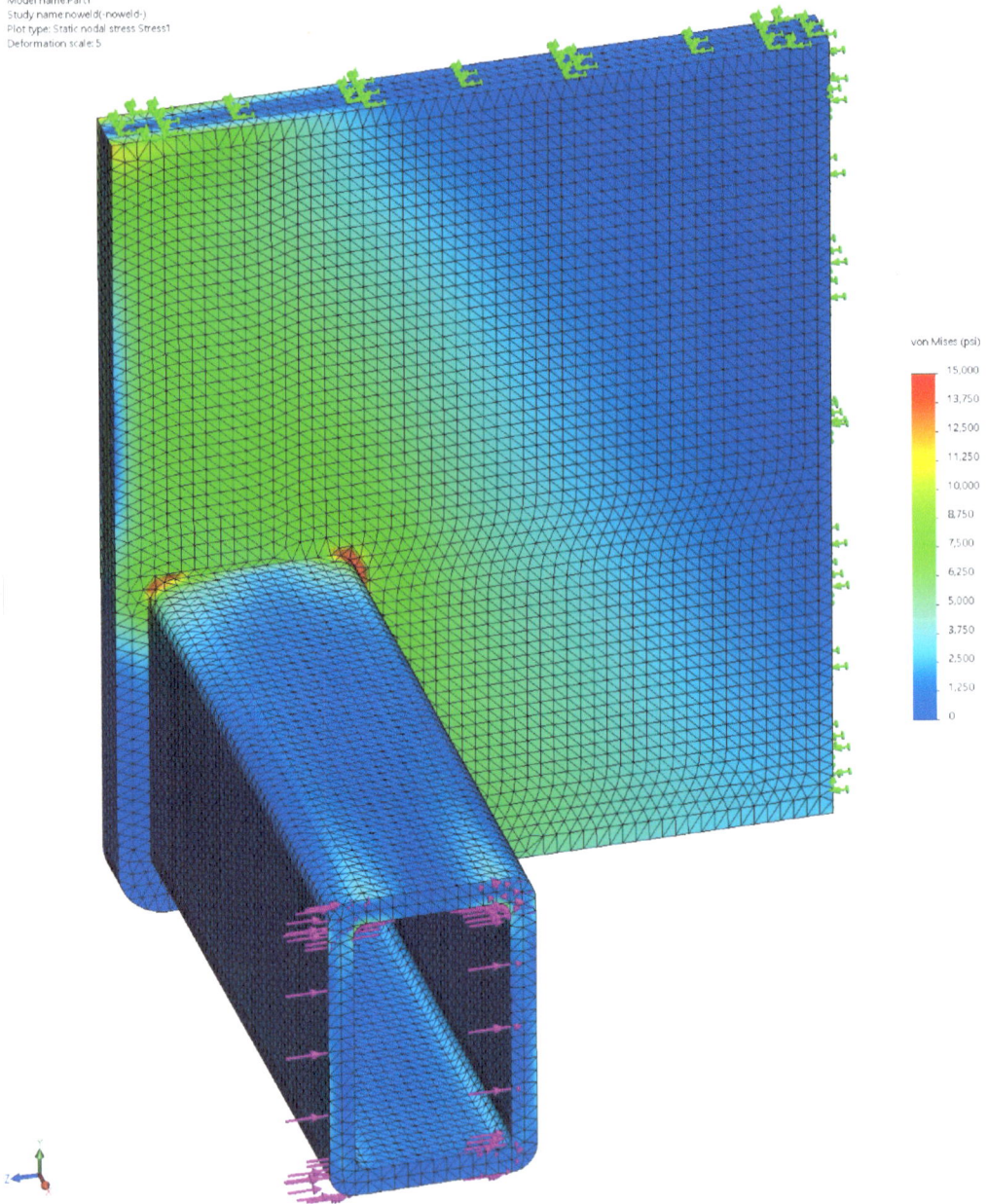

Model name:Part1
Study name:noweld(-noweld-)
Plot type: Static nodal stress Stress1
Deformation scale: 5

von Mises (psi)

15,000
13,750
12,500
11,250
10,000
8,750
7,500
6,250
5,000
3,750
2,500
1,250
0

First result, von Mises stress

The stress response is not surprising, hot spots appearing at the sharp corners.

But what about the welds? In this case it is easy to add the textbook weld shape with a single chamfer operation.

Model name:Part1
Study name:nocut(-nocut-)
Plot type: Static nodal stress Stress1
Deformation scale: 5

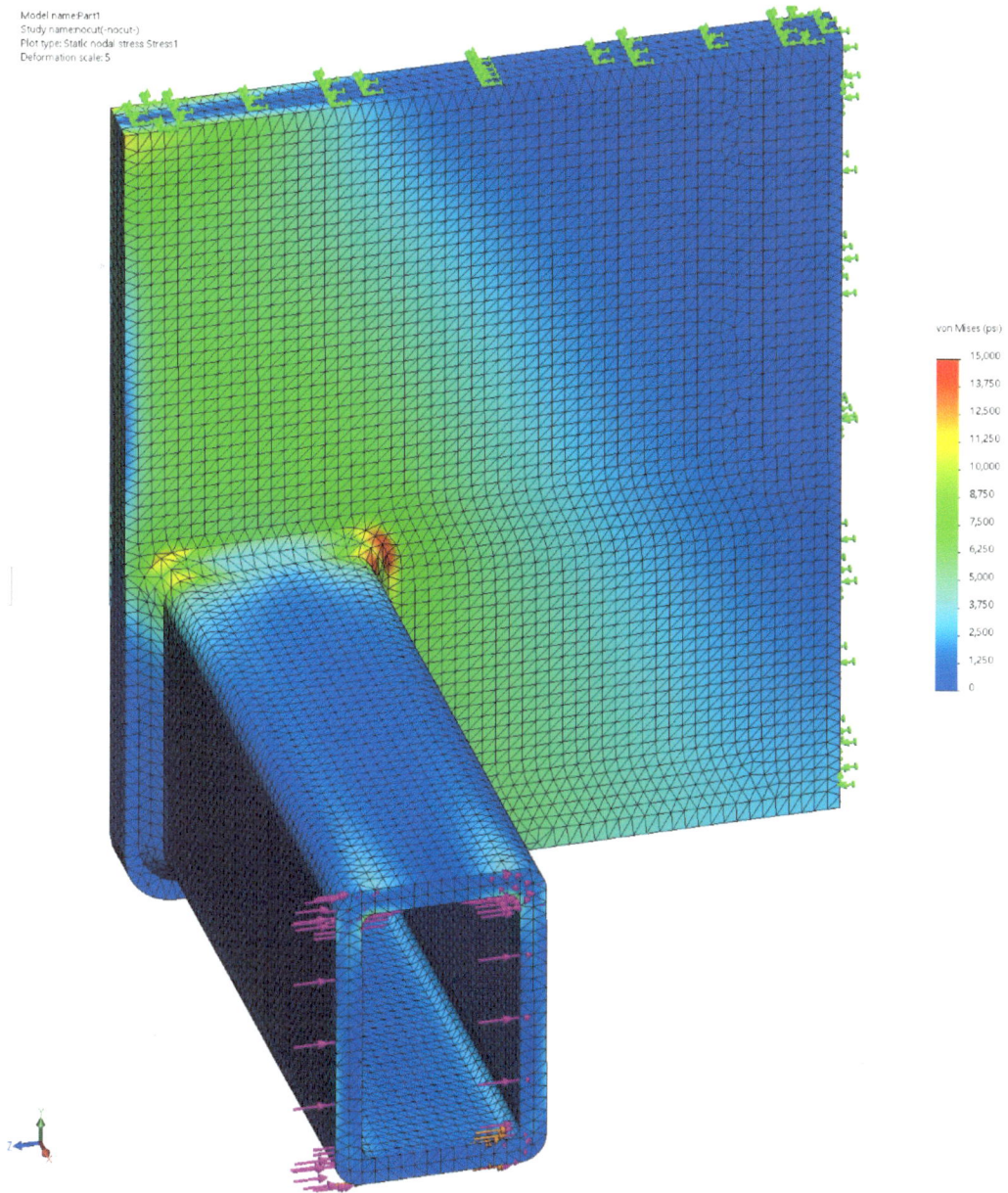

von Mises (psi)

15,000
13,750
12,500
11,250
10,000
8,750
7,500
6,250
5,000
3,750
2,500
1,250
0

Result with fillet weld added

The stress hot spots appear softer, which is not surprising as material was added and the sharp angles were increased. But is this still a conservative analysis? Probably not. Material is joined where there could not possibly be a bond.

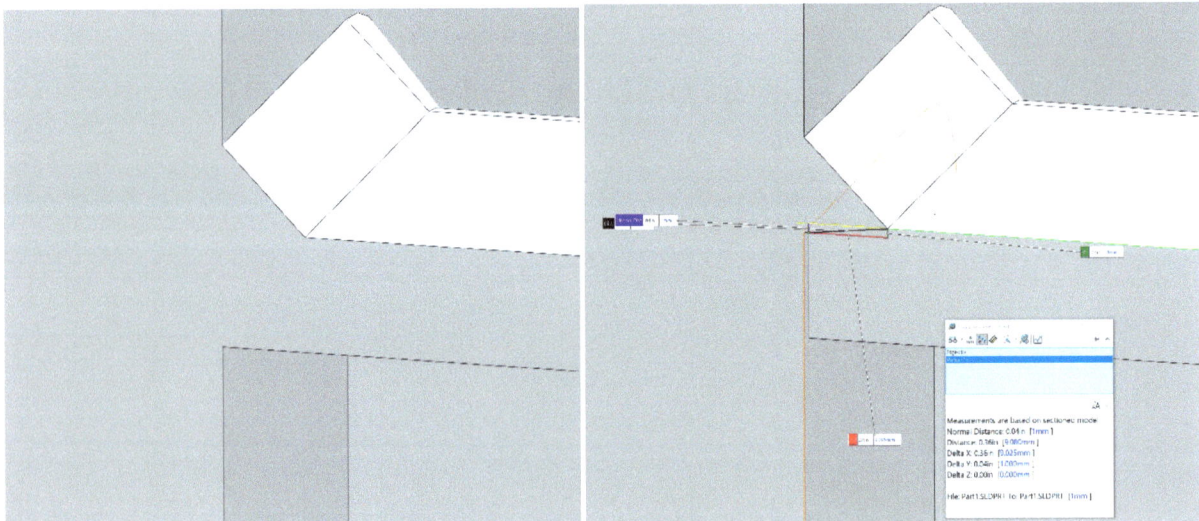
Geometry with weld ungapped vs. gapped

Gapping the Weld

Here two section views illustrate a change made for the next study. A thin gap has been cut between the tube and the plate. In Solidworks this was by an offset sketch of the tube perimeter. The offset was 1 mm, which the welder may have claimed is expected minimum penetration. The depth of the gap is 0.5 mm – more on that value later.

To put some numbers on the change, the cross-sectional area of the bond with the plate goes from 16.46 in^2 to 8.37 in^2, a 49% reduction.

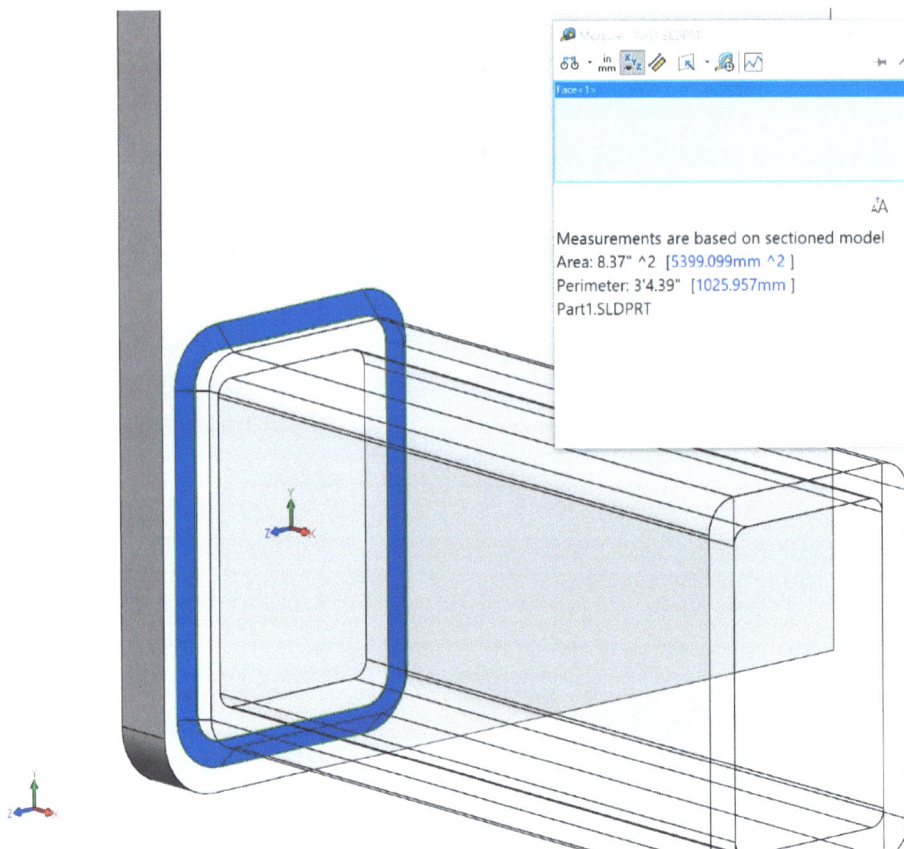
Section view through gapped solid model

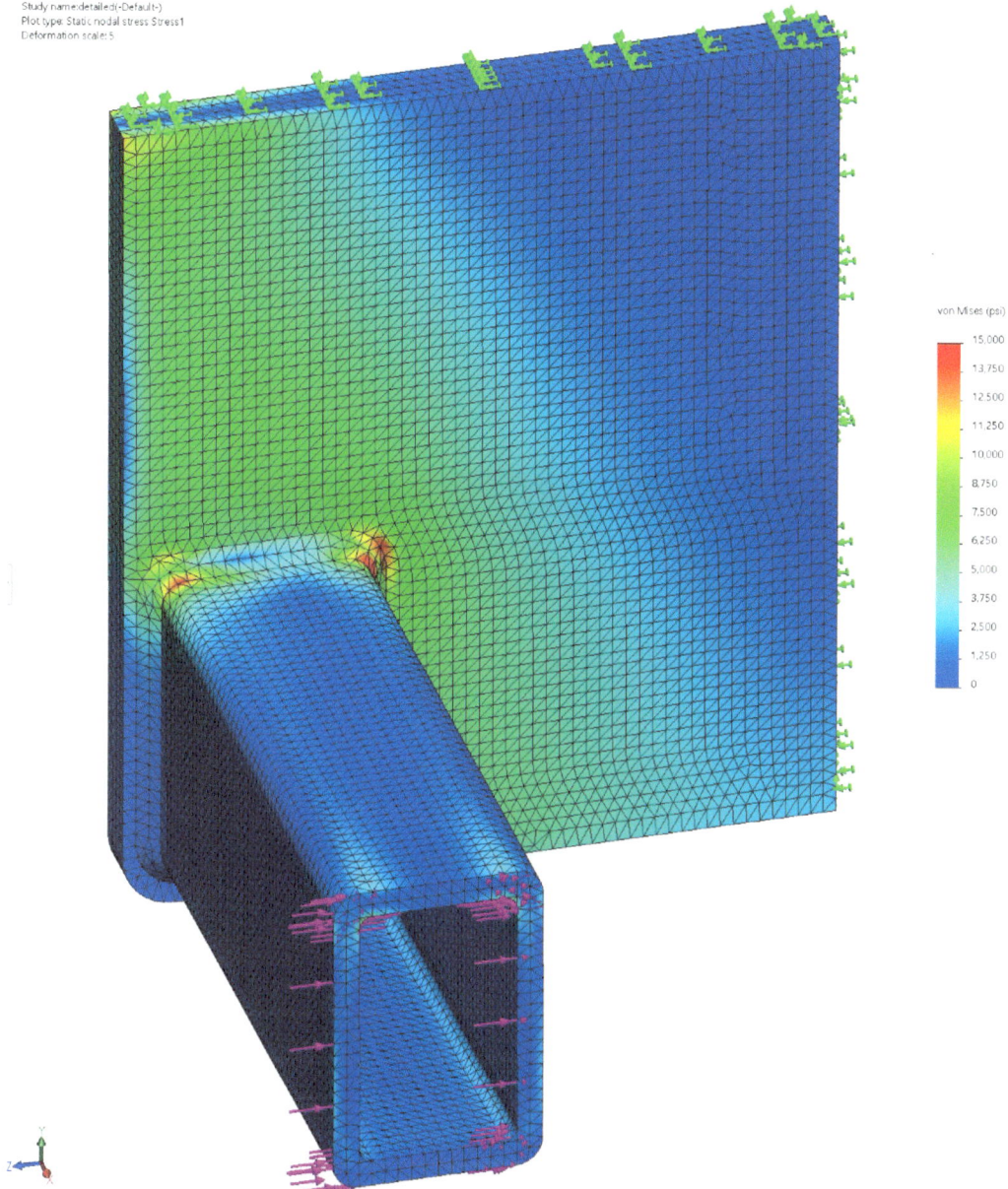

Model name:Part1
Study name:detailed(-Default-)
Plot type: Static nodal stress Stress1
Deformation scale:5

von Mises (psi)

15,000
13,750
12,500
11,250
10,000
8,750
7,500
6,250
5,000
3,750
2,500
1,250
0

Result with gapped model, von Mises stress

Very little material was removed in the gap cut, but with the reduced joint area surface stresses get appreciably hotter. We will talk more later about evaluating the stress peaks. The point made here is that a much more realistic representation of the final part geometry was studied at the cost of a very small additional effort.

Study 2: Crossmember Side Joint

The next model is a portion of a C-channel rail with a brace. It might be one of a pair of rails, with the brace being a crossmember between the pair. We have imposed a compound shear load on the cut surface of the crossmember, as if the rails were being pulled and sheared apart.

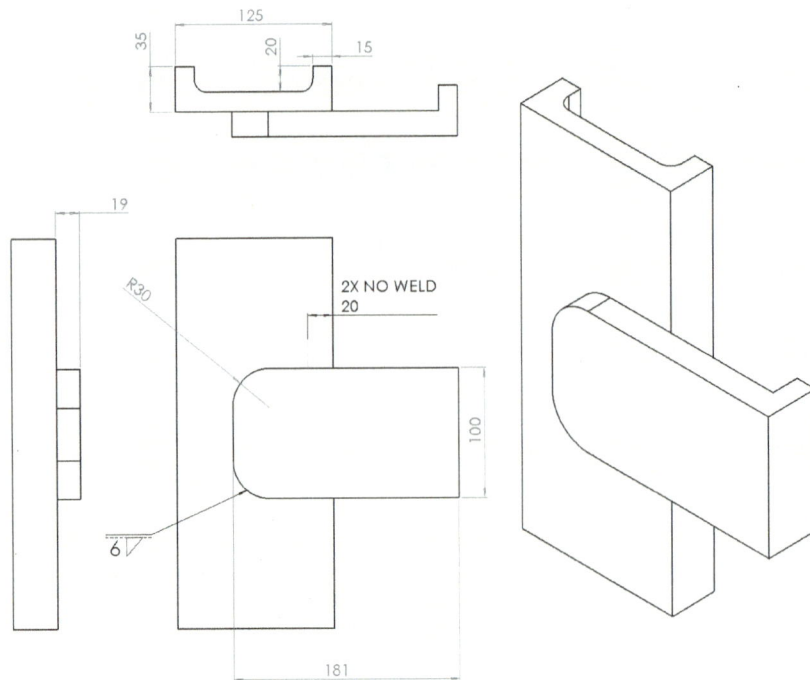

Drawing of weldment

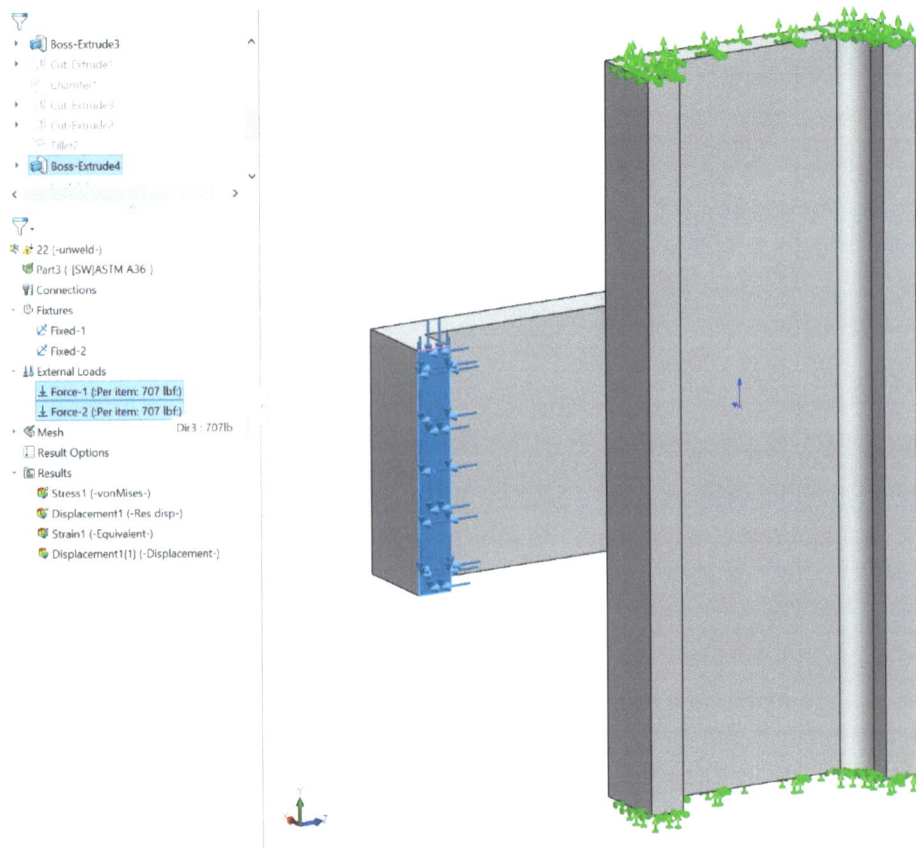

Simulation setup

At a first pass we can run the geometry as it comes, simply joining the bodies as one. The stress field is not very interesting.

Solidworks Simulation for Real Weldments

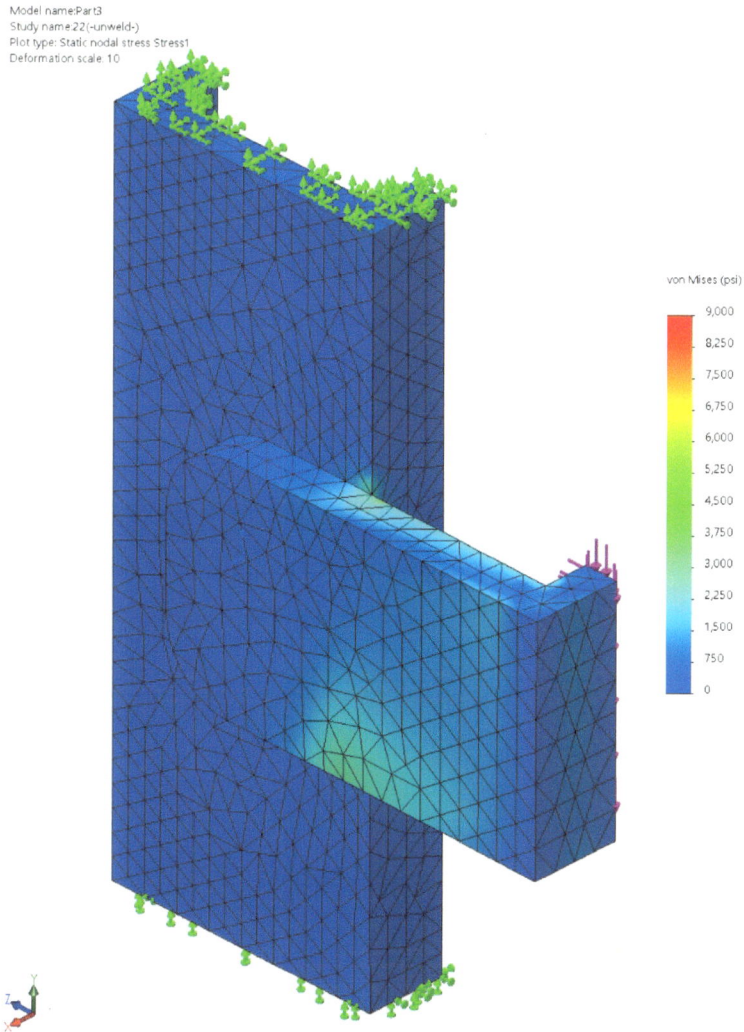

von Mises (psi)

```
9,000
8,250
7,500
6,750
6,000
5,250
4,500
3,750
3,000
2,250
1,500
750
0
```

Result, von Mises stress

We add the weld and put a mesh refinement on it. No gap is made behind it yet. Note that the drawing calls for the weld to be stopped short of the edge of the rail. This is common when beams or rails are loaded in bending, so as not to introduce a sudden change in section where the bending stresses are highest.

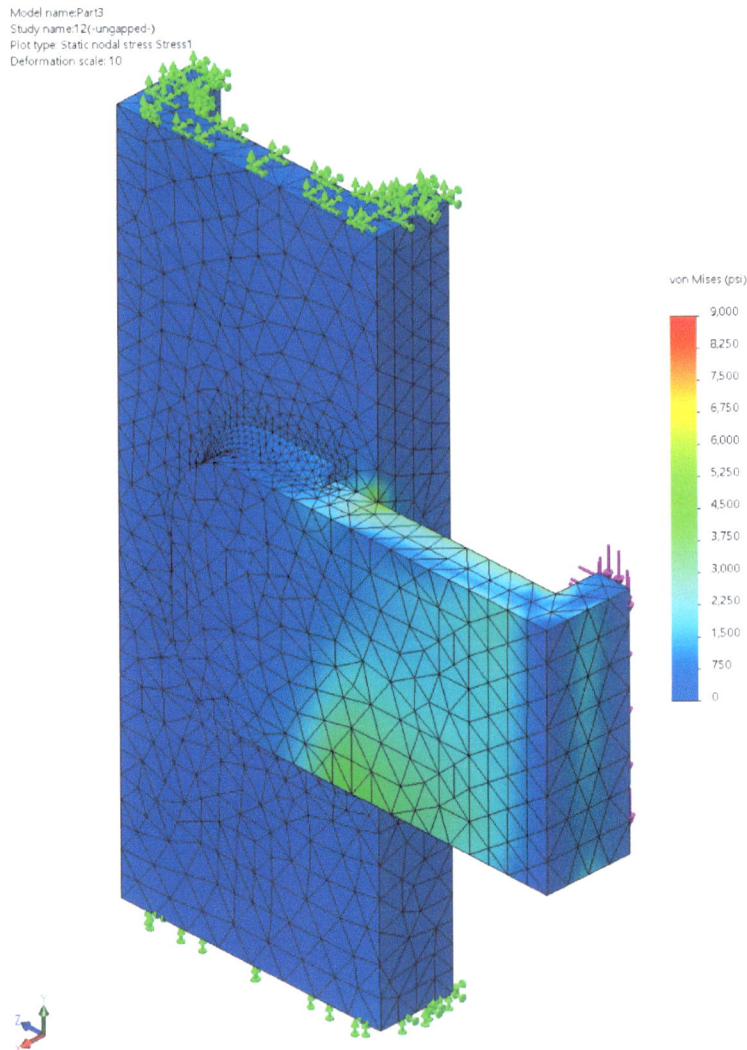

Model name:Part3
Study name:12(-ungapped-)
Plot type: Static nodal stress Stress1
Deformation scale: 10

von Mises (psi)

9,000
8,250
7,500
6,750
6,000
5,250
4,500
3,750
3,000
2,250
1,500
750
0

Result with weld added

No stress peaks are seen in the weld, but in this case the rail and brace plate are still joined on the entire mating surface. We must put a gap inside the weld bead.

Gapping the Partial Overlap

The gapping operation is more complicated here. A cut is made offset from the weld as before, but now it must be carried out some distance away from the end of the overlap between the parts. For this model a generous fillet is used to blend the cut back into the full thickness part of the plate.

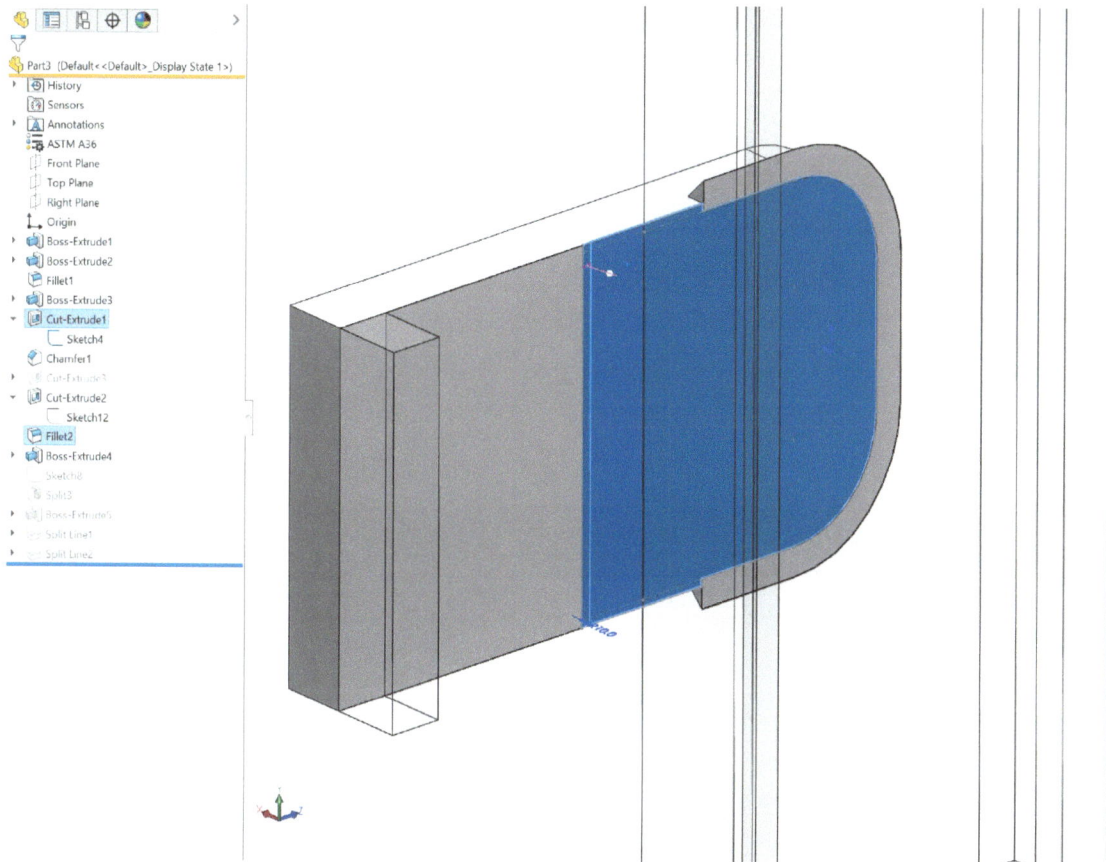

Section view of model with gap cut

A quick note on modeling with chamfer operations – they don't always behave as desired. In this case the chamfer will overrun the intended limits and continue to the extents of the side plate.

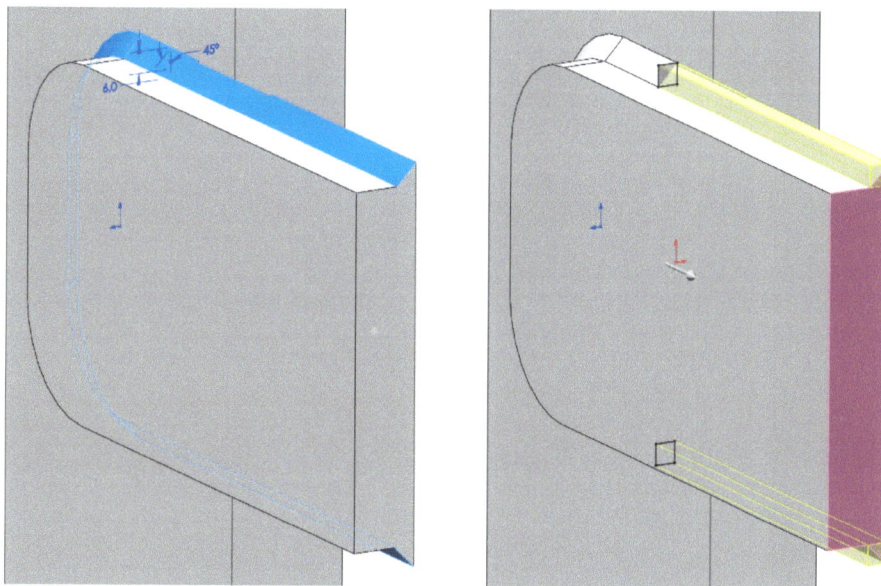

Model with chamfer overrun and cut-extrude to correct

Here it is not a big job to make a single cut-extrude operation to trim off the excess material.

When cutting the gap we gave it a discrete value, well above zero, in this case 0.5 mm. This is about $1/40^{th}$ of the part thickness, $1/12^{th}$ of the weld size, and (most importantly) about $1/8^{th}$ of the element size used on the weld bead. The sizing is important to avoid severely high aspect ratios.

Mesh section through aspect ratio plot

An aspect ratio of about 7 is no problem at all to run. We could make the gap thinner, even infinitesimal. But aspect ratios in the hundreds, or mesh failures, or a lot of effort splitting and joining bodies are the eventual consequences. We have found the geometric approximation of an appreciable gap to be a much smaller detriment to accuracy than leaving parts whole and un-gapped.

Another note on mesh generation – sometimes it may be helpful to add mesh refinement on the inside of the weld, on the narrow gap surface. This will ensure multiple elements across the thickness of the welds (where high stress gradients may exist), reduce aspect ratios, and improve odds of meshing success.

We run the analysis and find high stress at the end of the weld bead.

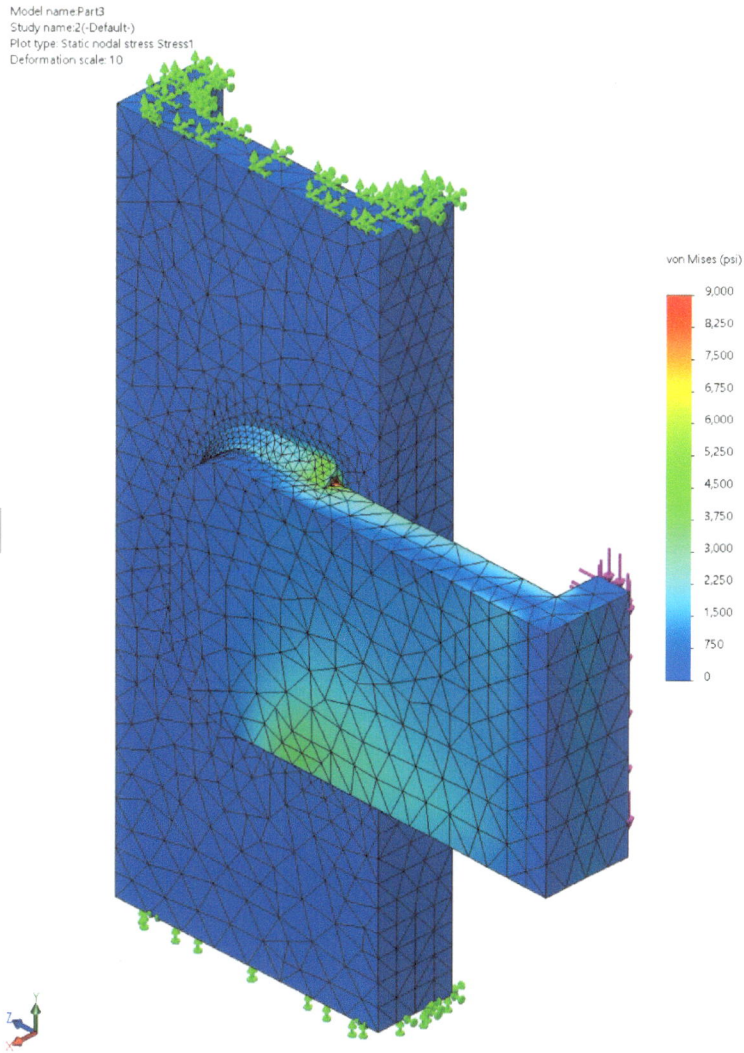

Model name:Part3
Study name:2(-Default-)
Plot type: Static nodal stress Stress1
Deformation scale: 10

von Mises (psi)

9,000
8,250
7,500
6,750
6,000
5,250
4,500
3,750
3,000
2,250
1,500
750
0

Result with gapped fillet weld

The stress hot spot is of obvious interest, so a finer mesh refinement is put on the weld, the study is re-run, and we zoom in on the hot spot, looking for an exact stress value. Here we hit the hard limits of finite element technology.

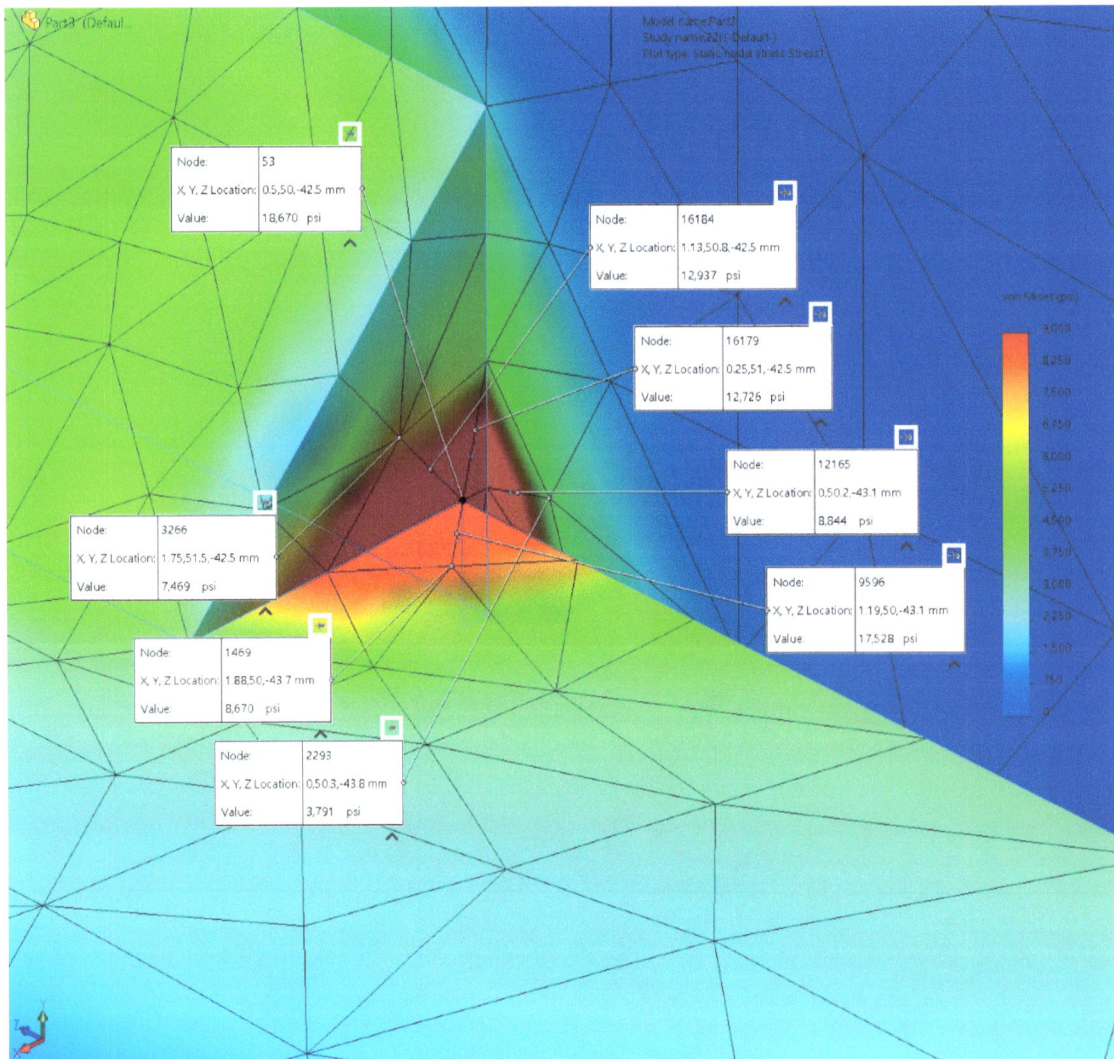

Stress concentration at end of weld bead

FEA Results on Sharp Corners

There is nothing fundamentally wrong with FEA methods. If the code is developed rigorously, FEA can well represent continuum mechanics on any geometry. The problems here are:
 - Continuum mechanics cannot handle perfectly sharp edges, and
 - We don't really know what the exact geometry is going to be.

Other sources cover the limitations of continuum mechanics when confronted with a perfect sharp. In such a location stress calculations will always tend toward infinity, and an FEA result will follow that trend as the mesh gets finer.

As for the true weld geometry, even the cleanest fillet bead has some blending at the molten edges at best, or bad undercut at worst. At the end of the weld if not specified it may taper off, or end abruptly as in this model, or anything in between. The abrupt sharp end is probably worst case for stress peaks, but what stress value here do we use in design?

After much experience with physical prototypes (and torn strain gages), we have some guidelines. In this example the triple point in the middle, showing 18670 psi, is certainly to be discarded. Using Solidworks' default 2nd order tetrahedral elements we can look at the adjacent mid-nodes, or the element corner nodes one element away from the sharp geometric corner. The mid-nodes get up to 17,528 psi (von Mises stress); the corner nodes peak at 8670 psi. The true peak lies somewhere between the last two numbers This is not a very helpful situation, values a factor of two apart, but the design itself can be adjusted to avoid sharp edges being the peak stress zones.

Improved Joint Design

An alternative design for the crossmember is seen below. This design has several advantages, including having a clean consistent end location of the weld bead and the ability to use a smaller crossbar with lower peak stress.

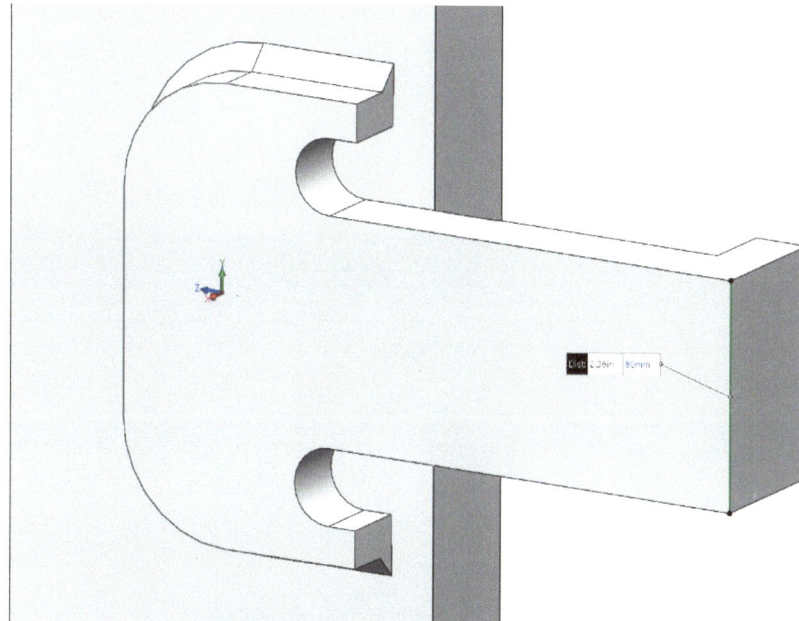

Model with "T" shaped joint

The inside radius of the "T" shape becomes a tune-able control of the peak stress made by the load transfer. In the following result the weld is modeled and a gap cut behind it as before.

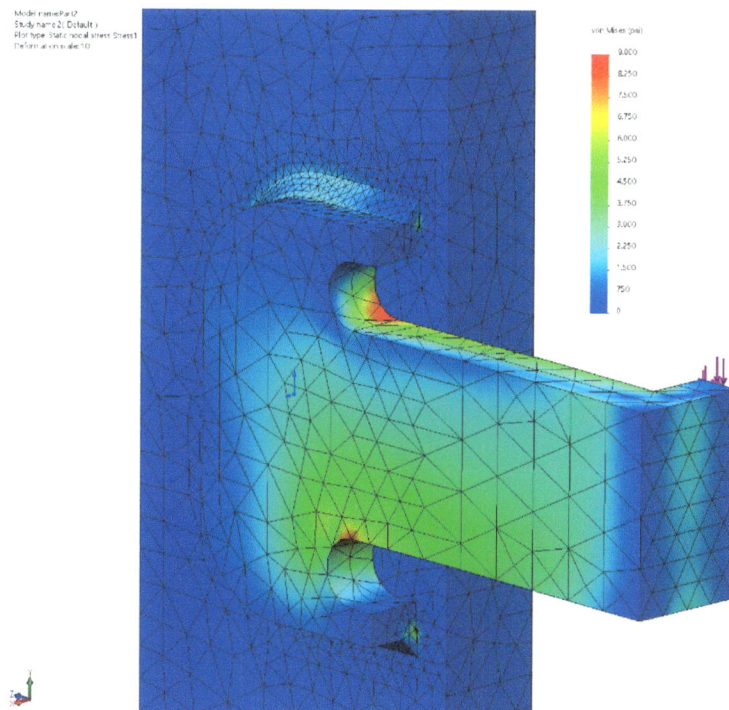

Result, von Mises stress

We can probe a value of about 10,500 psi on the radius and have high confidence that this is a near exact solution.

Modeling Separate Solids

Up to now we have modeled each weldment as a single solid with uniform properties. If it should be necessary to have different properties in the parts or the weld filler, we need only make them separate solids. We start with a merged part as before, as this allows creation of the weld bead with a chamfer operation. Then a series of splits and one combine operation get us to having three solid bodies (in intimate contact).

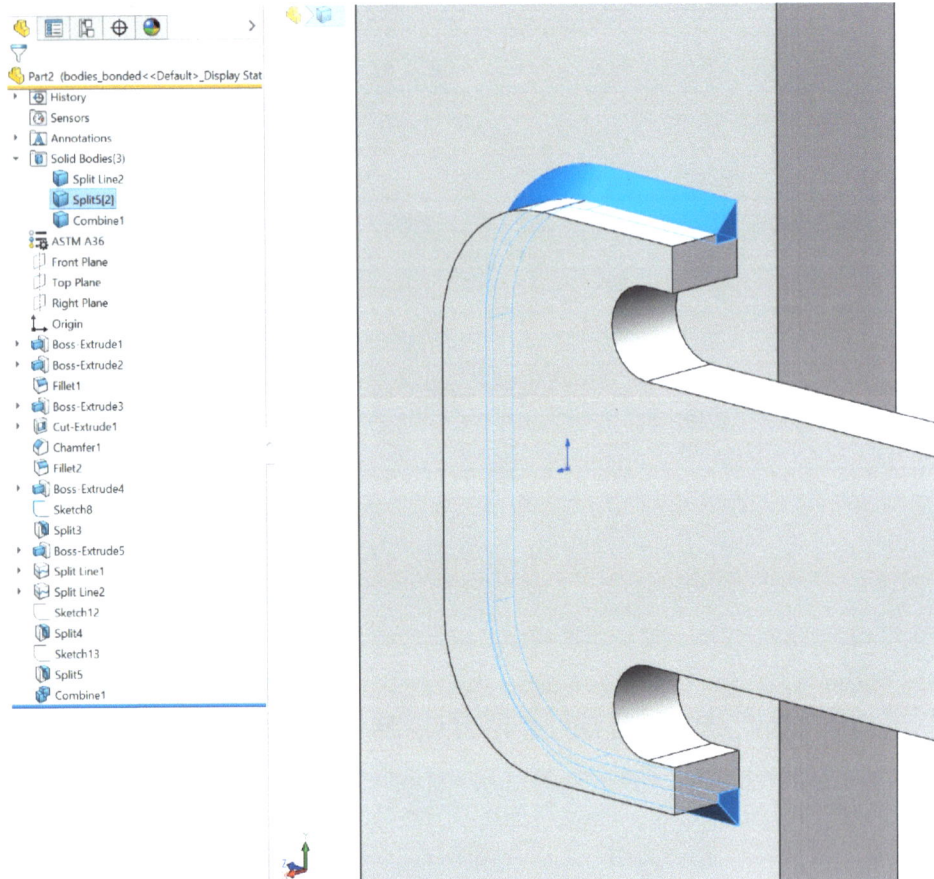

Model split into three solids

It would be nice to use global bonded contact in Solidworks Simulation, but this would make the gapped area bonded as well. So three bonded contacts are made manually (this involves some hiding of bodies and/or some clever filtering).

Simulation setup, contacts highlighted

Note also that "compatible mesh" is not an option in manual bonded contact. The best we can do is refine all the mating surfaces and activate higher accuracy incompatible bonding.

For good measure in this run a no penetration contact is added between the T-plate and the rail surface. It is possible for them to press on each other.

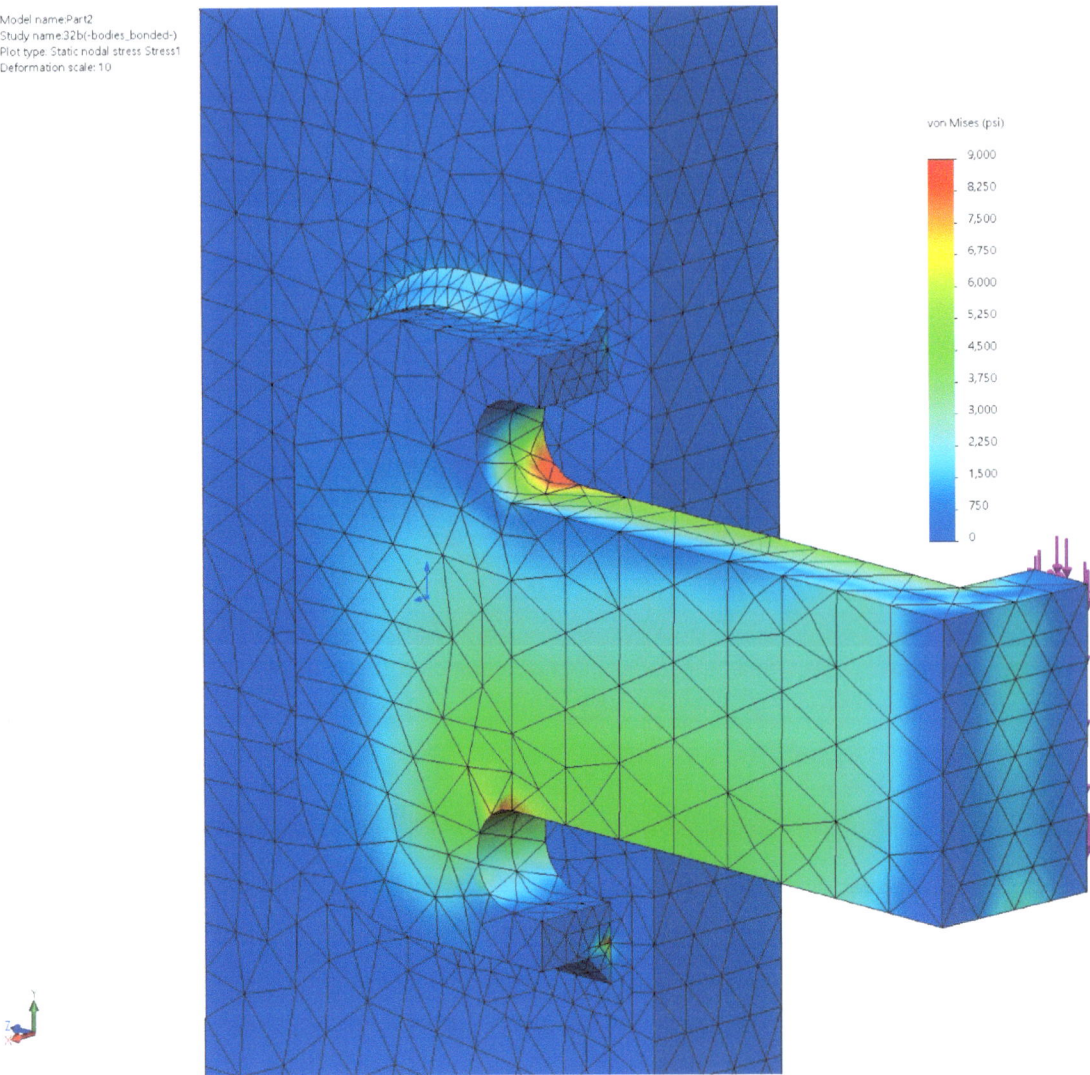

Model name:Part2
Study name:32b(-bodies_bonded-)
Plot type: Static nodal stress Stress1
Deformation scale: 10

von Mises (psi)

9,000
8,250
7,500
6,750
6,000
5,250
4,500
3,750
3,000
2,250
1,500
750
0

Result with multi-body model

This method is the most laborious, of course, but besides the option of different materials it offers the ability to get force and component stress values from all the new mating surfaces.

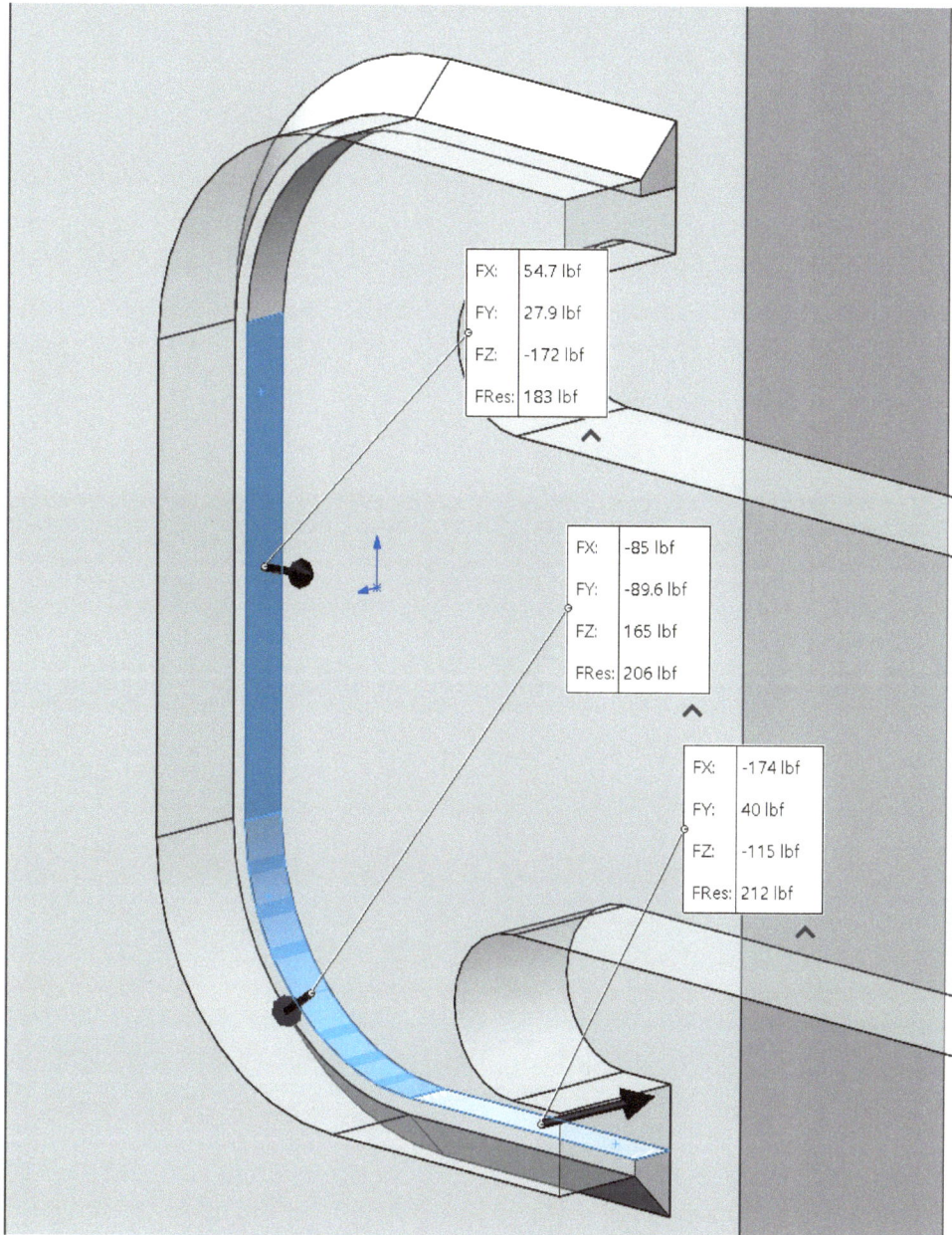

FX:	54.7 lbf
FY:	27.9 lbf
FZ:	-172 lbf
FRes:	183 lbf

FX:	-85 lbf
FY:	-89.6 lbf
FZ:	165 lbf
FRes:	206 lbf

FX:	-174 lbf
FY:	40 lbf
FZ:	-115 lbf
FRes:	212 lbf

Interface force values at faces of modeled weld penetration

Study 3: Lift Frame

Now we get to something that looks almost like a real machine. This weldment supports a load cantilevered out a distance from a support frame. The frame may be installed in a larger assembly which moves the whole thing around in any direction.

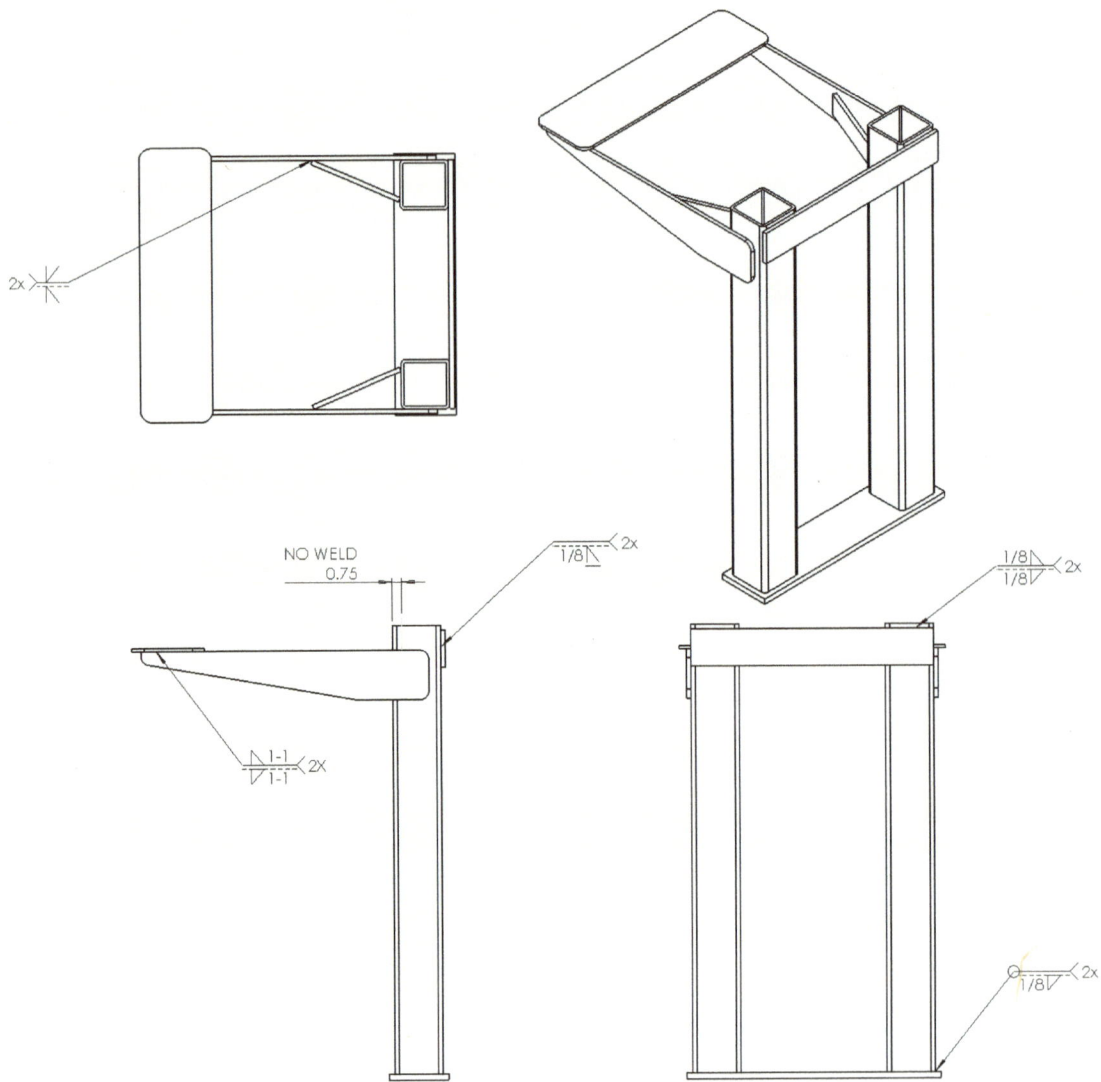

Drawing of weldment

The weldment consists of A513 tube steel posts with all other parts A36 plate. It's about three feet high with a 21 inch reach. The bottom plate is attached to the larger assembly and may be taken as the simulation fixture.

Solidworks Simulation for Real Weldments

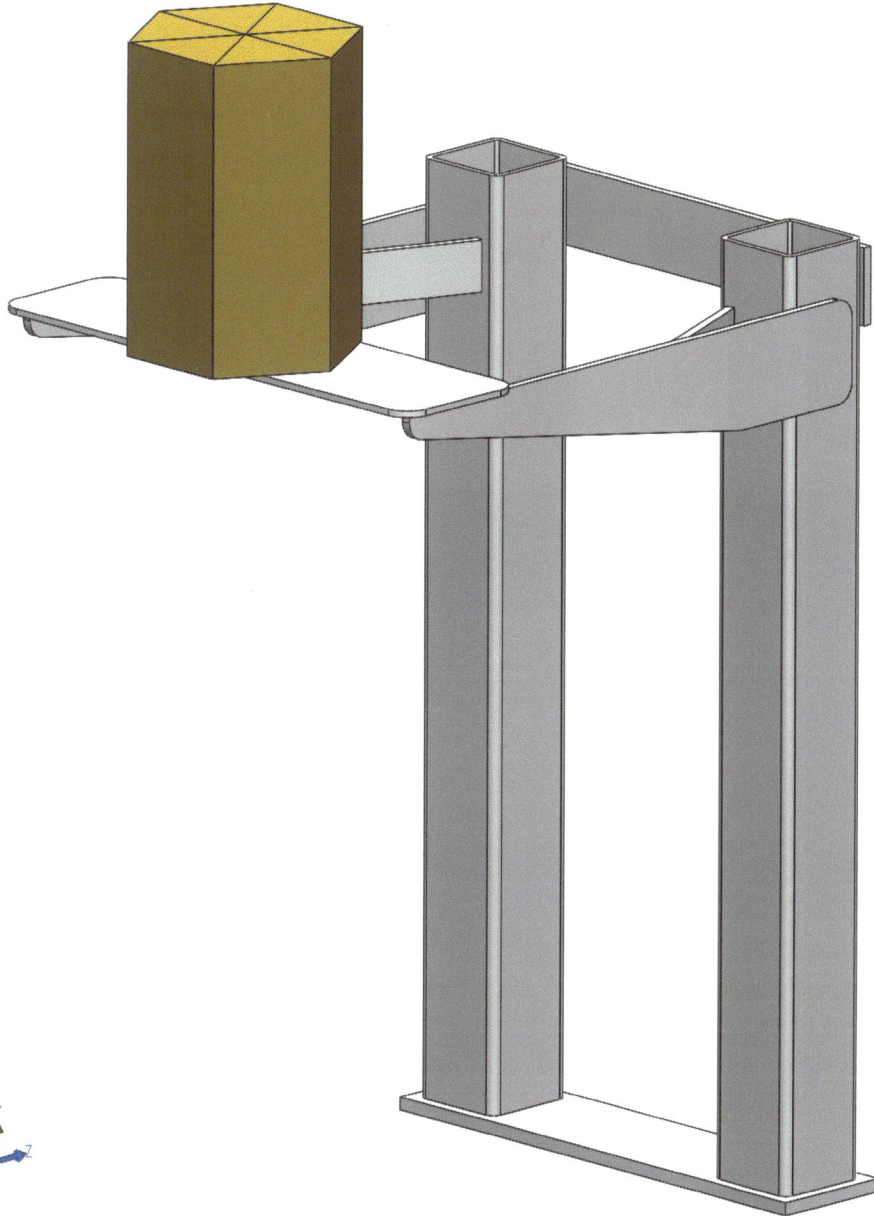

Model with load

The load here, a section of eight inch brass hex, is modeled explicitly. When we go to set up the analyses, acceleration (from gravity and motion) will be the only input. There is no guessing about the proper input forces, and no typing errors in them. {Of course, inertial loading also requires correct mass be assigned to each material!]

The assembly loads into simulation and meshes on the first try with no complications. It is tempting to simply leave "global bonded contact" turned on, add a gravity vector and one constraint, and run it.

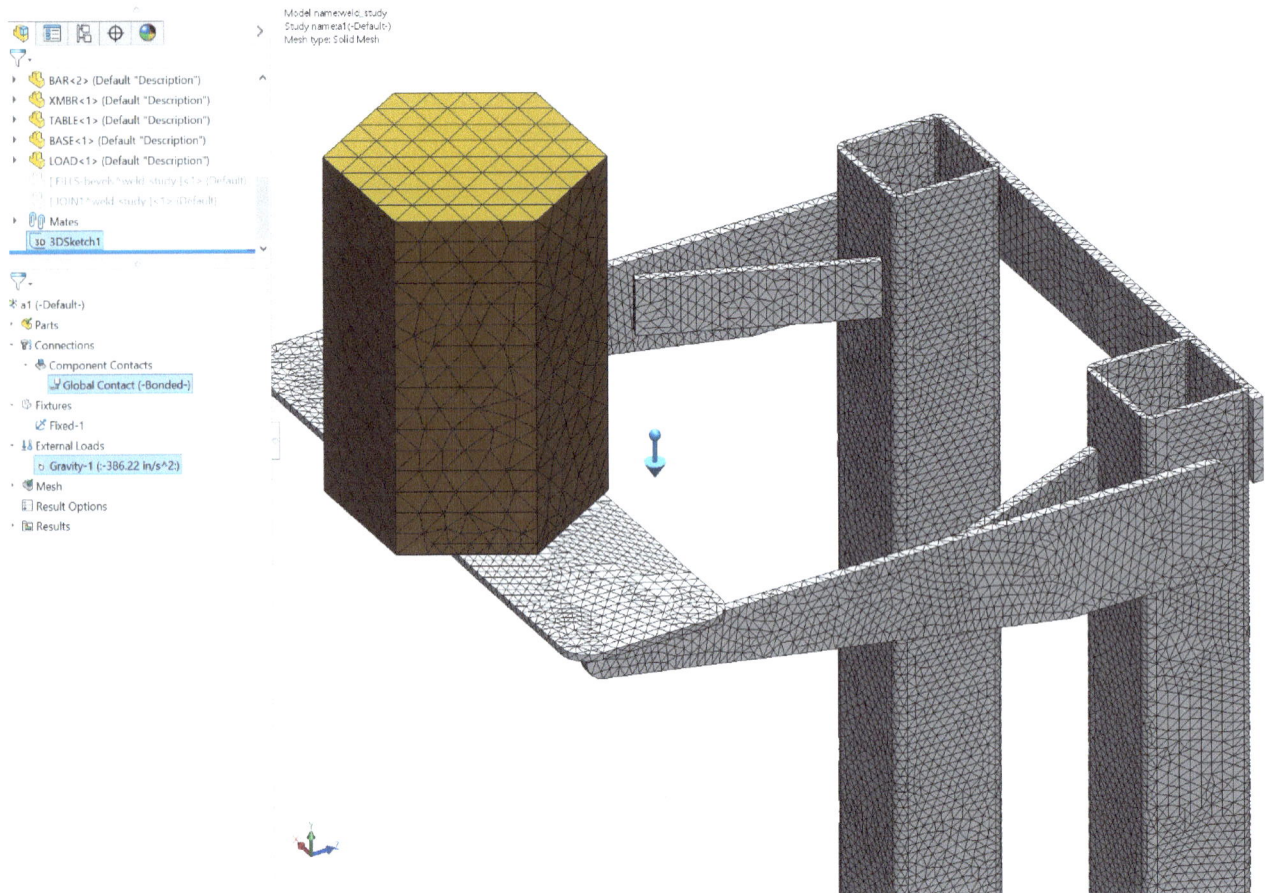

Model name:weld_study
Study name:a1 (-Default-)
Mesh type: Solid Mesh

BAR<2> (Default "Description")
XMBR<1> (Default "Description")
TABLE<1> (Default "Description")
BASE<1> (Default "Description")
LOAD<1> (Default "Description")
Mates
3DSketch1

a1 (-Default-)
Parts
Connections
Component Contacts
Global Contact (-Bonded-)
Fixtures
Fixed-1
External Loads
Gravity-1 (:-386.22 in/s^2:)
Mesh
Result Options
Results

Initial simulation study setup and mesh

There is one big problem with this approach – the diagonal braces are not connected and they bear no forces. In the assembly model they are mated edge-on-face. It also looks like the table has stress hot spots, but note the low scale limit and hold that thought for later.

Model name:weld_study
Study name:a1 (-Default-)
Plot type: Static nodal stress Stress1

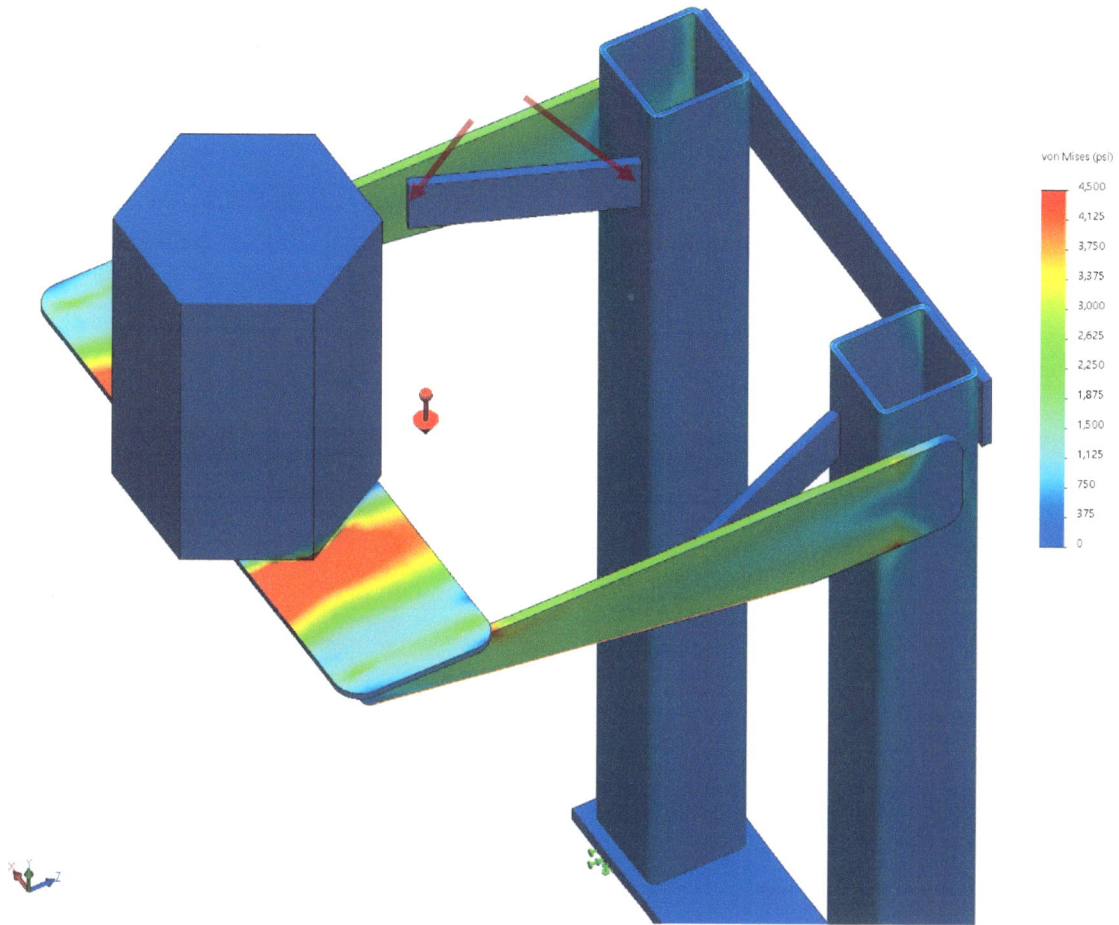

Stress result on first study

The braces are completely un-stressed and the arms take all the bending load. If a deformed plot is made the braces will be off-screen, as they were unrestrained from motion.

As usual we will try to be very literal in the model to solve the problem. A new in-context part is added to the assembly. As the braces are identical and positioned symmetrically a single sketch-extrude operation fills the volume of the four weld beads at the brace ends.

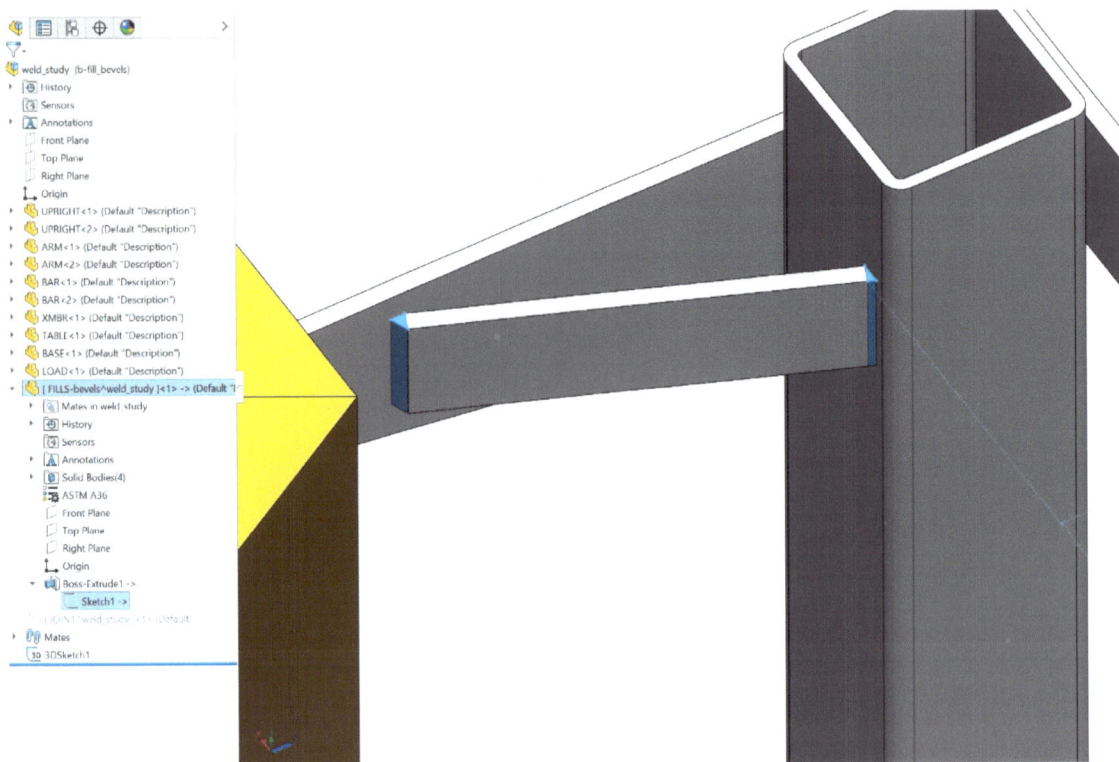
Model with weld filler added

Now the braces do take some load and the stress pattern in the long arms is generally cooler inward of the brace connections.

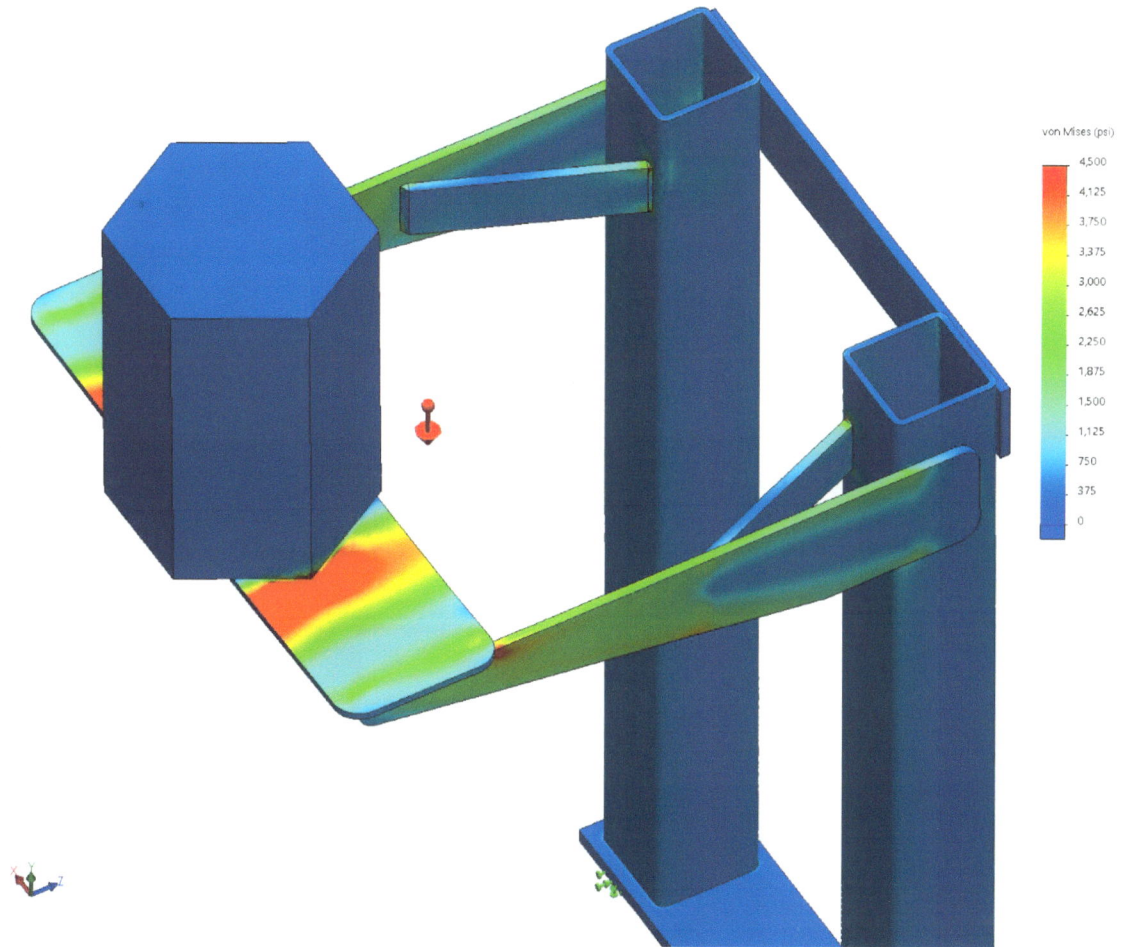

von Mises (psi)

4,500
4,125
3,750
3,375
3,000
2,625
2,250
1,875
1,500
1,125
750
375
0

Stress on study with added weld bead part

Incompatible Meshes

In the global bonded contact we left 'compatible mesh' selected. This costs a few more elements in the study, so what if we left it off?

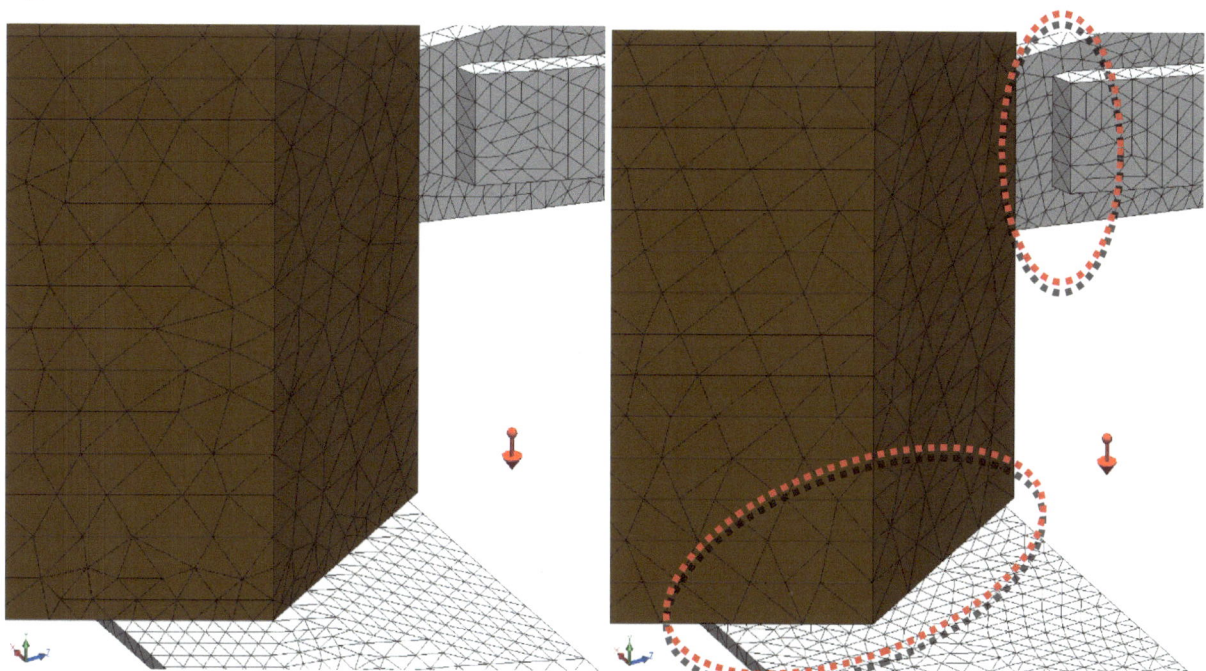

Mesh with compatible and incompatible mesh at bonded contact

If one zooms in changes in the stress pattern can be seen. As the new peaks and dips are at the joints, we can't just ignore them and would rather not have to explain them away. We'd like to use compatible mesh – but remember that this is (so far) only an option for *global* bonded contact.

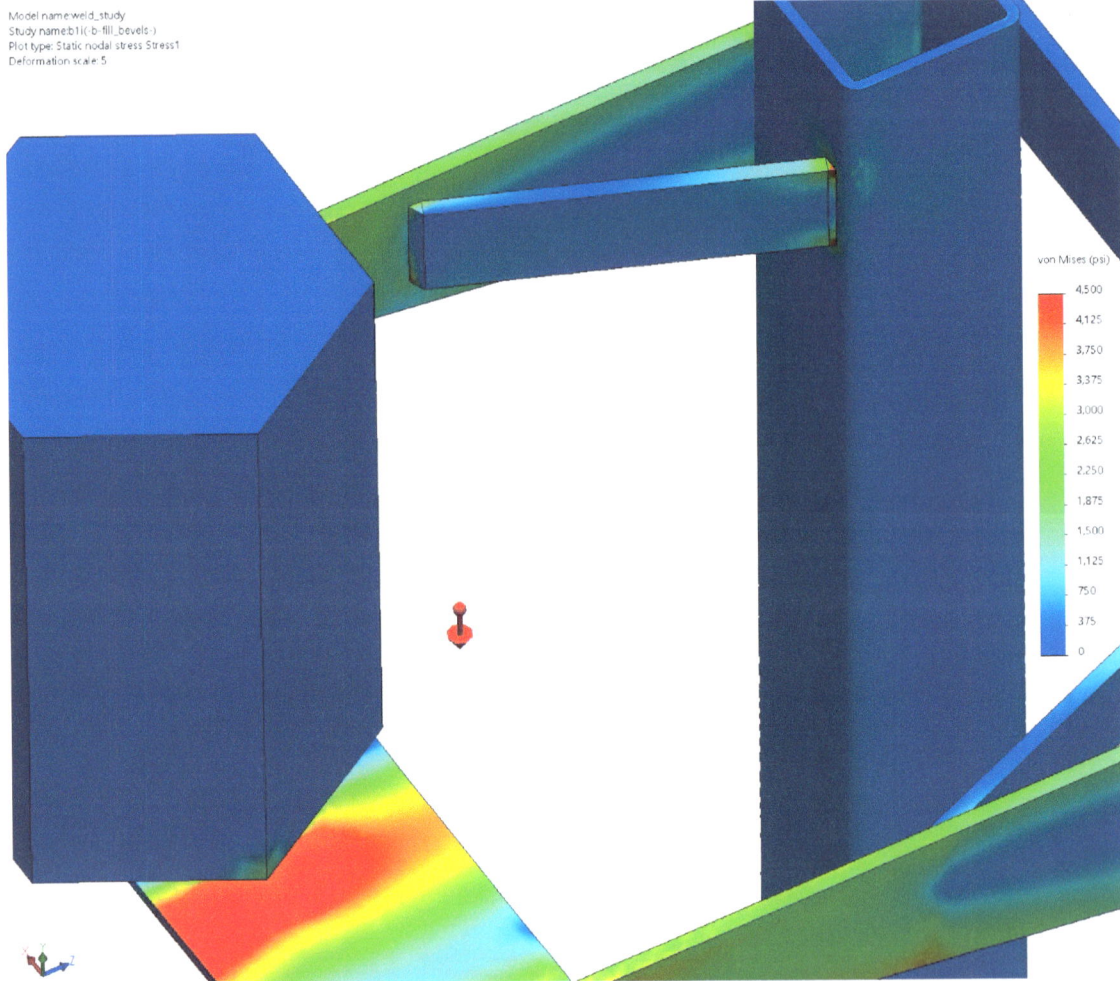

Model name:weld_study
Study name:b1(-b-fill_bevels-)
Plot type: Static nodal stress Stress1
Deformation scale: 5

von Mises (psi)

4,500
4,125
3,750
3,375
3,000
2,625
2,250
1,875
1,500
1,125
750
375
0

Result with incompatible mesh at bonded contact

The other welds in the model will be more difficult to make as separate bodies. So we're going to switch back to making a single solid body.

Solid Body in Assembly

The previous weldment models were actually single part models. In this case we have a real assembly. It's usually most convenient to work with a single solid, but we probably want to leave the design model alone. So either in a configuration of the assembly or a new parent assembly we make a new in-context part.

The first operation is to join all the other parts into the root feature of the new part.

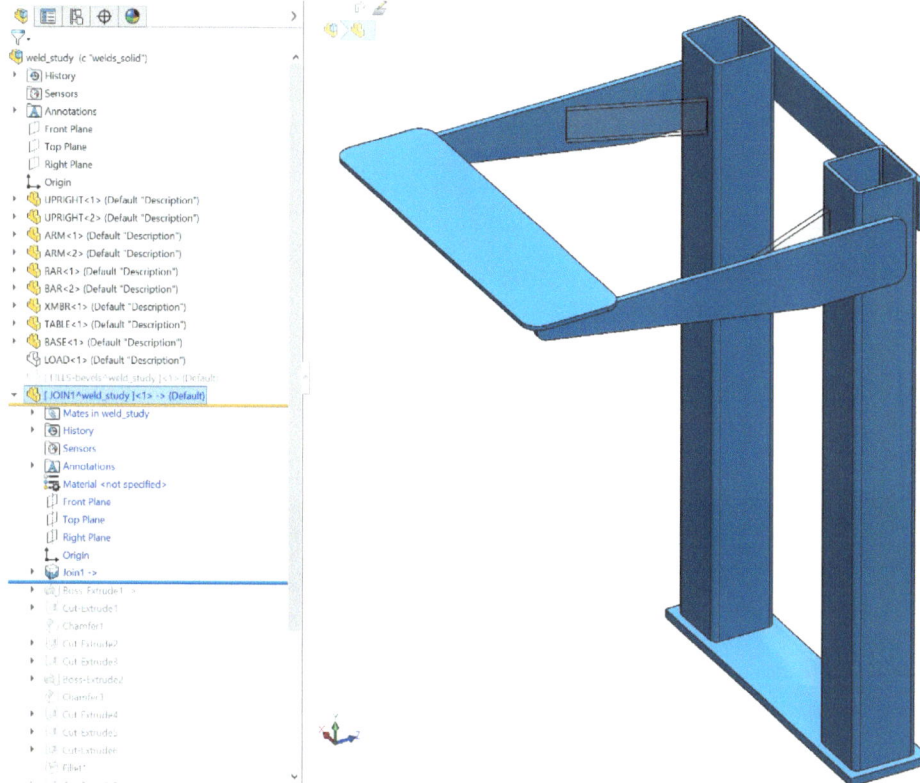

New solid root join feature

The diagonal braces are not included because they meet the other bodies edge-on-face. This does not make a solid body so is invalid for a "join" operation. We add the braces with a separate extrude operation (taking advantage of sketching in context) and add the weld filler at the same time.

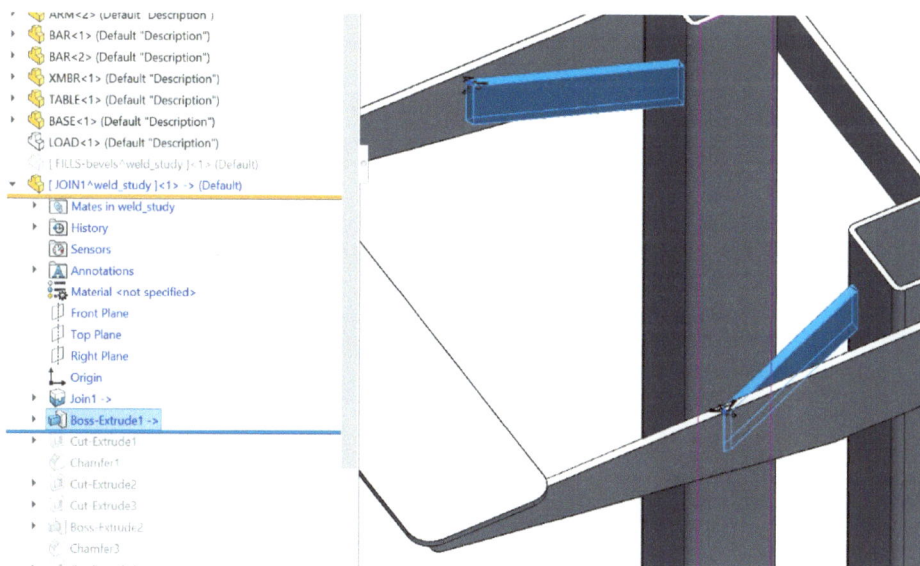

Extrude operation to add braces with welds

Editing the Standalone Solid Part

Now we can open the new part separately to work on it. First the gapped welds at the bottom of the tubes are made.

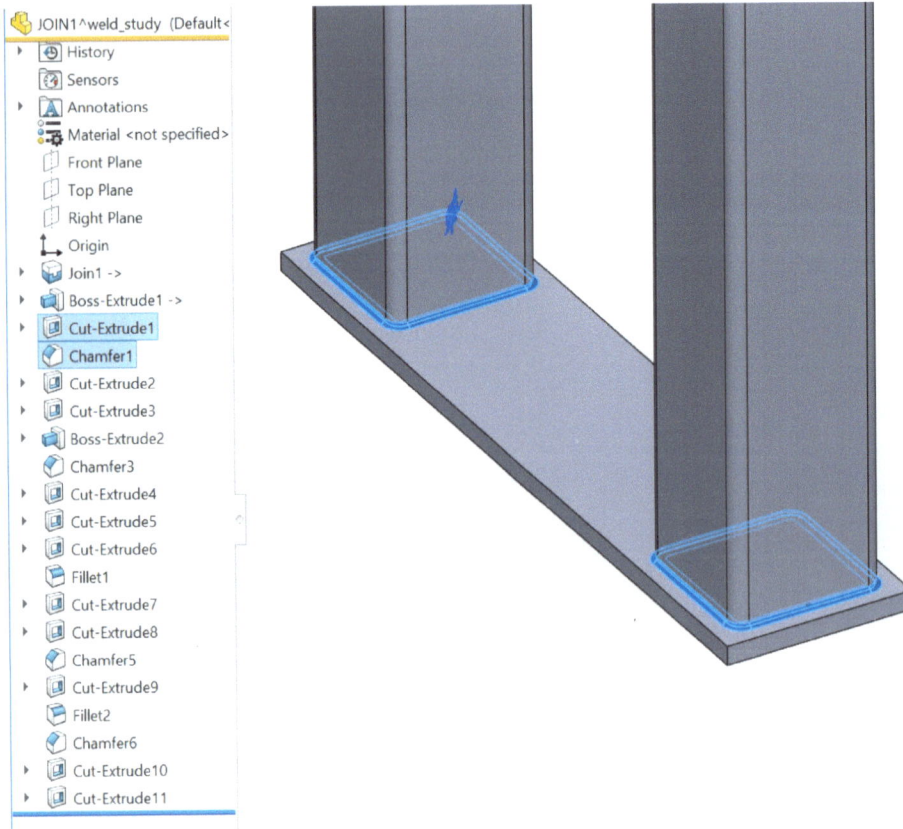

Weld operations for tube bottoms

The back plate is welded on three sides of each end.

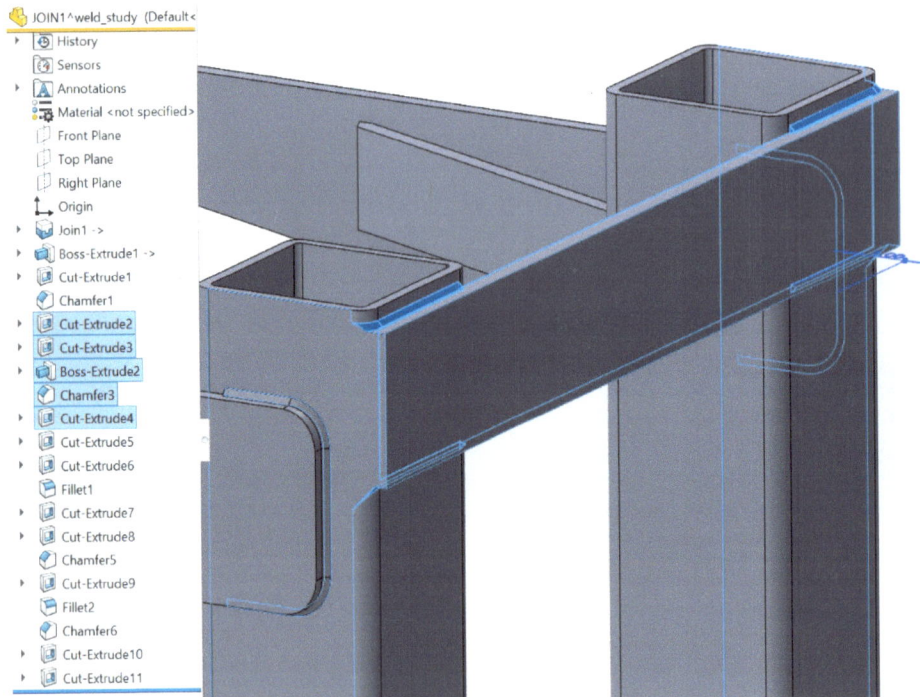

Operations for back plate welds

The gapping operation makes the entire place a sliver thinner. It is supposed that the top and bottom welds stop a bit short of the tube inside radius.

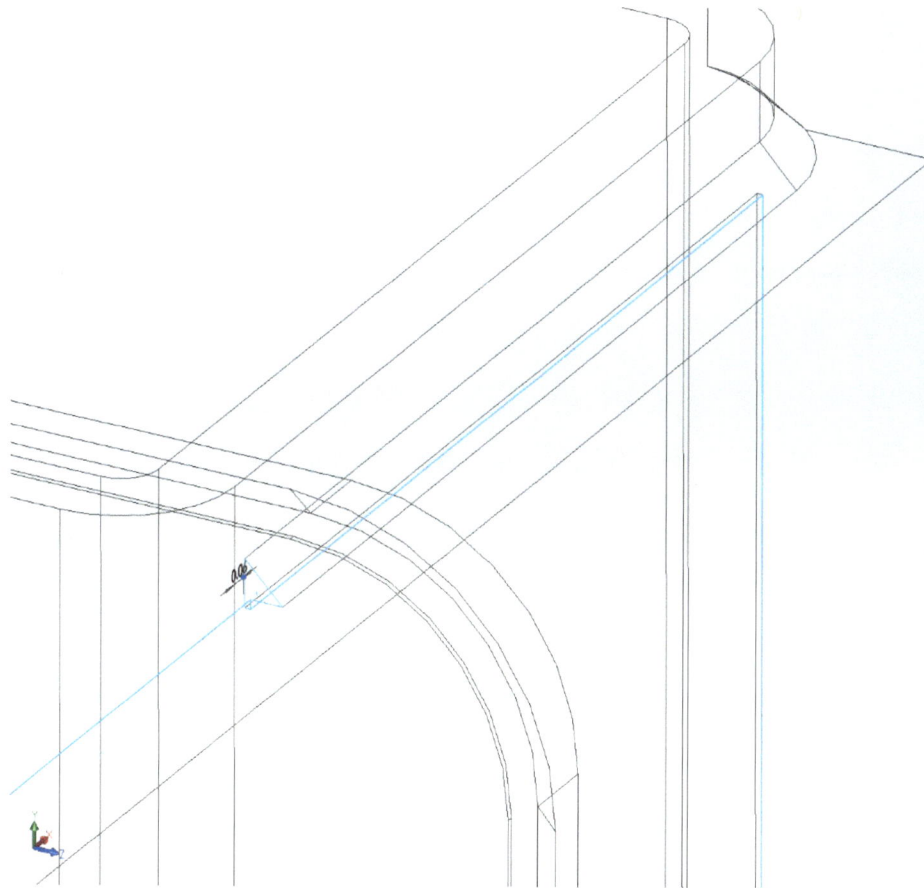

Detail of back plate weld

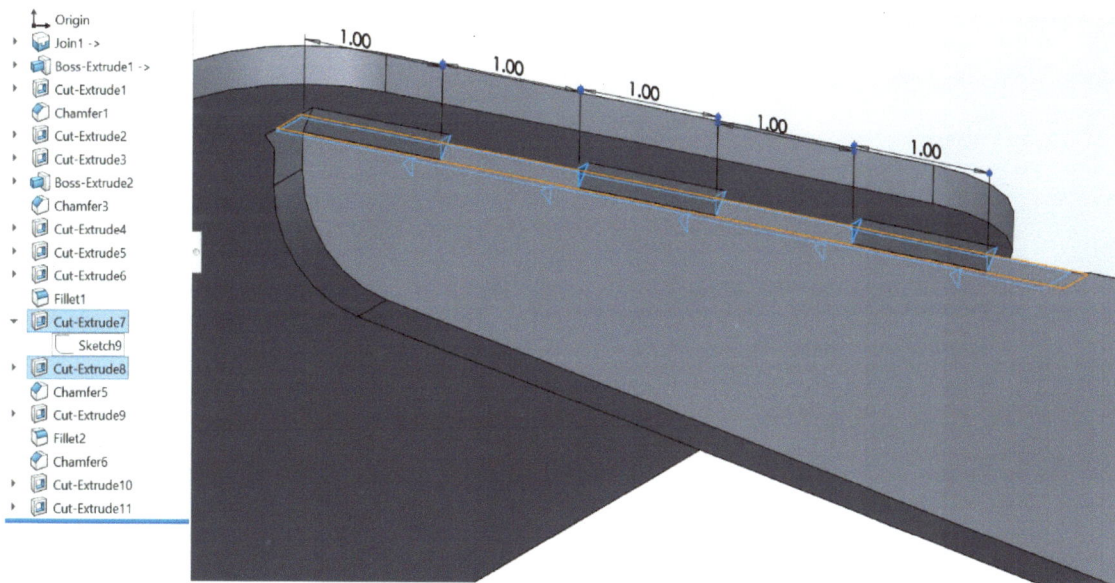

Table-arm staggered welds

Multiple operations are required to make the staggered welds for the table. Note that they are gapped, by cutting into the tops of the arms.

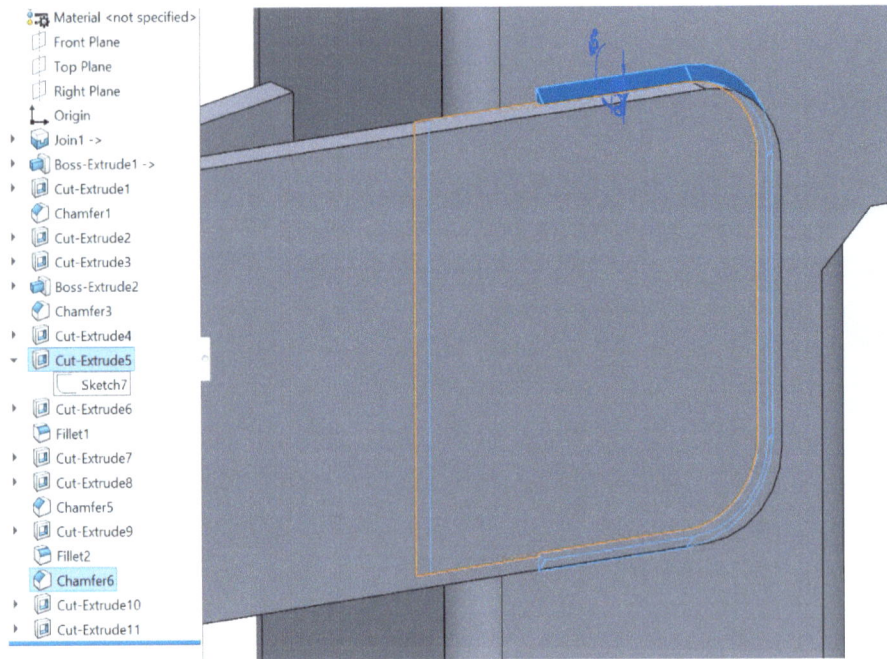

Gapped weld at rear of arm

New Study in Assembly

Now the model is complete. We can set up analyses at the part level or in the assembly. We stick with the assembly so as to copy settings over. The unused source parts are 'excluded' from the new study. [One can also break links in the join part then suppress all the source parts.]

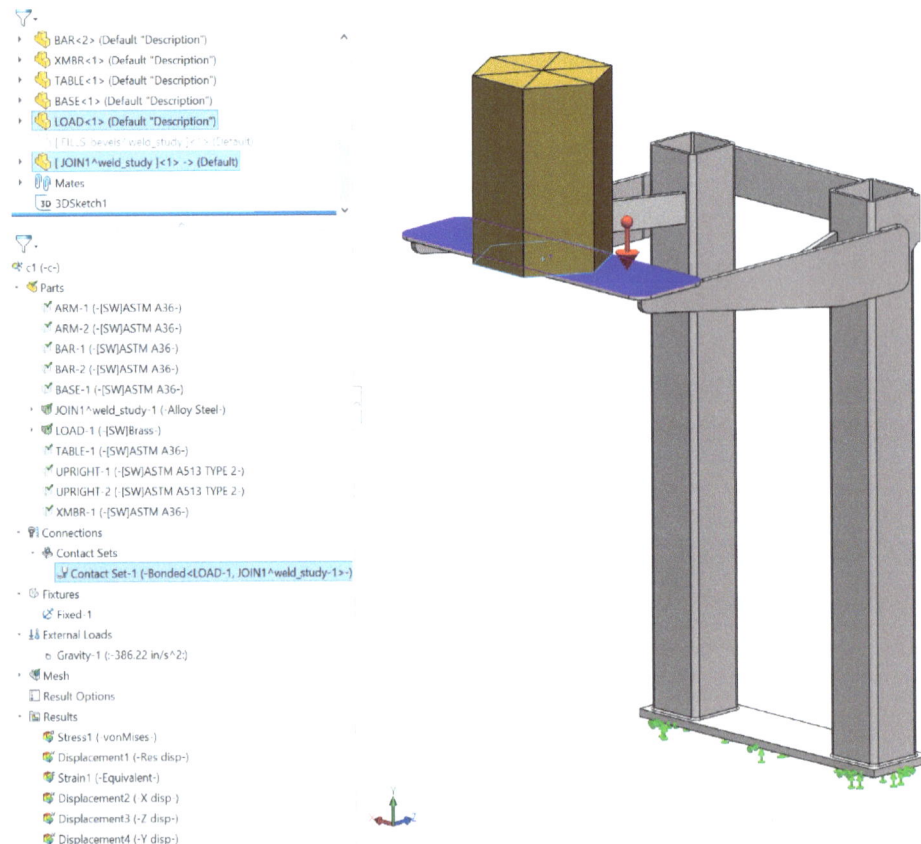

Study setup in assembly with joined solid part

Model name:weld_study
Study name:c1 (-c-)
Plot type: Static nodal stress Stress1
Deformation scale: 15

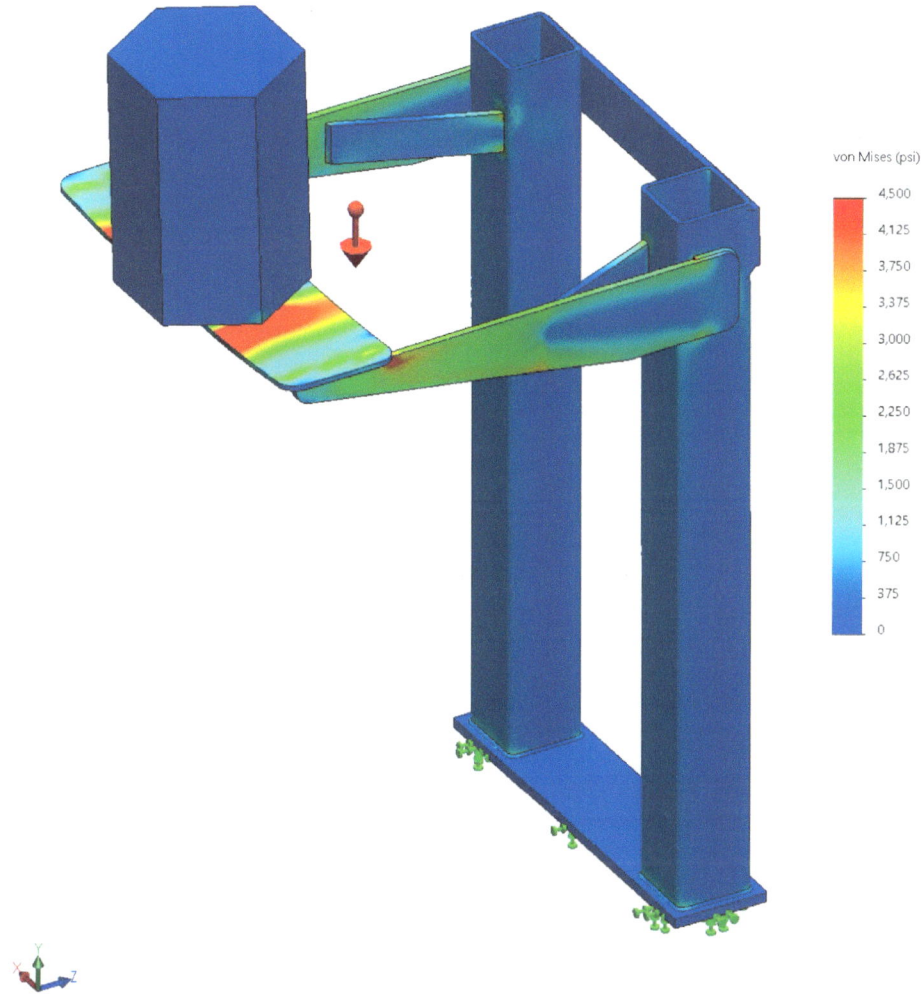

von Mises (psi)

4,500
4,125
3,750
3,375
3,000
2,625
2,250
1,875
1,500
1,125
750
375
0

Result on joined part, bonded contact with load

This study was run with the load bonded to the table. The table is a plate which might be too thin. If the table bends significantly, the bottom of the load will not entirely be in contact with it. With bonded contact on, the load will artificially stiffen the thin plate.

Accurate Contact for the Input Load

Another study is set up with a no penetration contact (with friction, to keep the load from flying off).

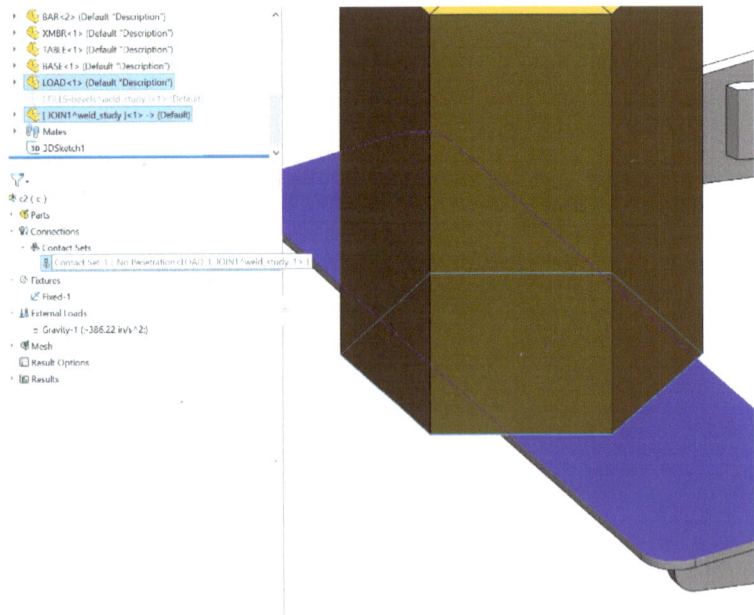

Contact condition

The change in contact has a big impact on stress and deformation of the thin plate.

Model name:weld_study
Study name:c1(-c-)
Plot type: Static nodal stress Stress1
Deformation scale: 15

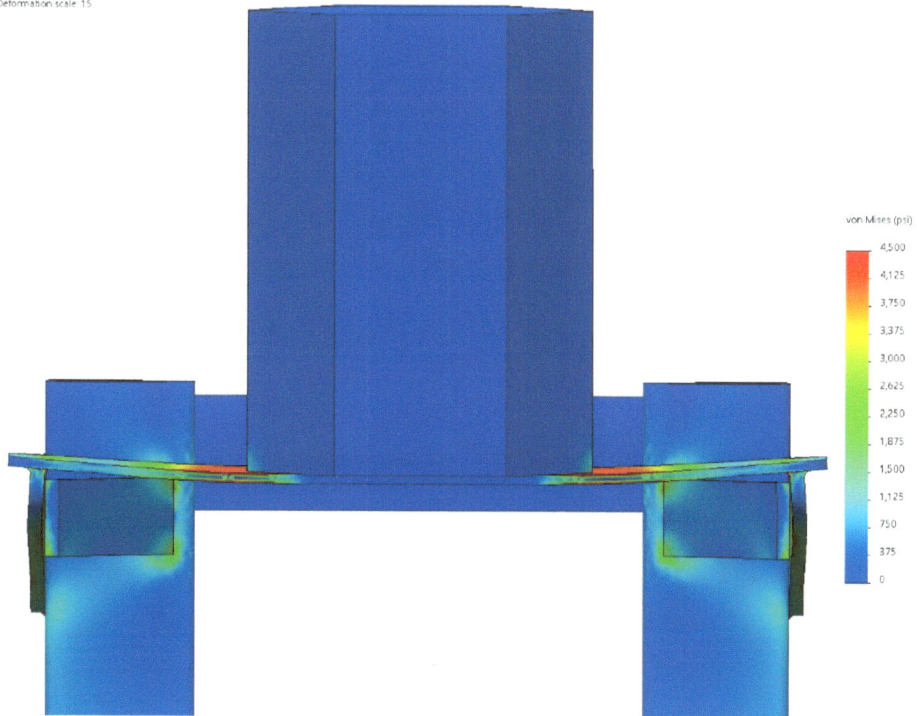

von Mises (psi)

4,500
4,125
3,750
3,375
3,000
2,625
2,250
1,875
1,500
1,125
750
375
0

Model name:weld_study
Study name:c2(-c-)
Plot type: Static nodal stress Stress1
Deformation scale: 15

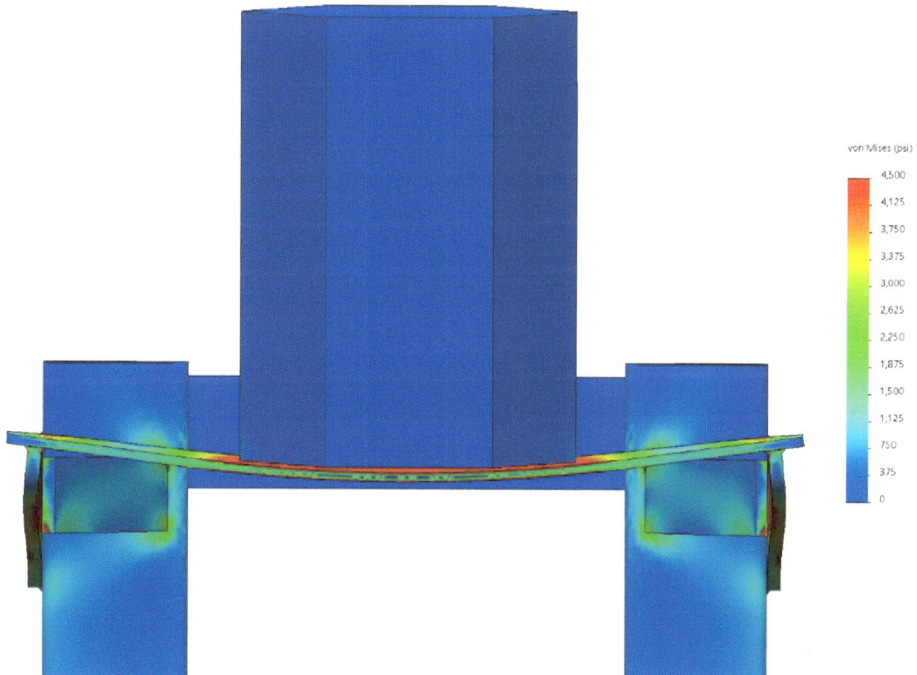

von Mises (psi)

4,500
4,125
3,750
3,375
3,000
2,625
2,250
1,875
1,500
1,125
750
375
0

Results with bonded and no penetration contact (15x magnified displacement)

The input of twisting forces into the arms is much greater in the latter study.

Model name: weld_study
Study name: x2(-:-)
Plot type: Static nodal stress Stress2
Deformation scale: 15

von Mises (psi)

6,000
5,500
5,000
4,500
4,000
3,500
3,000
2,500
2,000
1,500
1,000
500
0

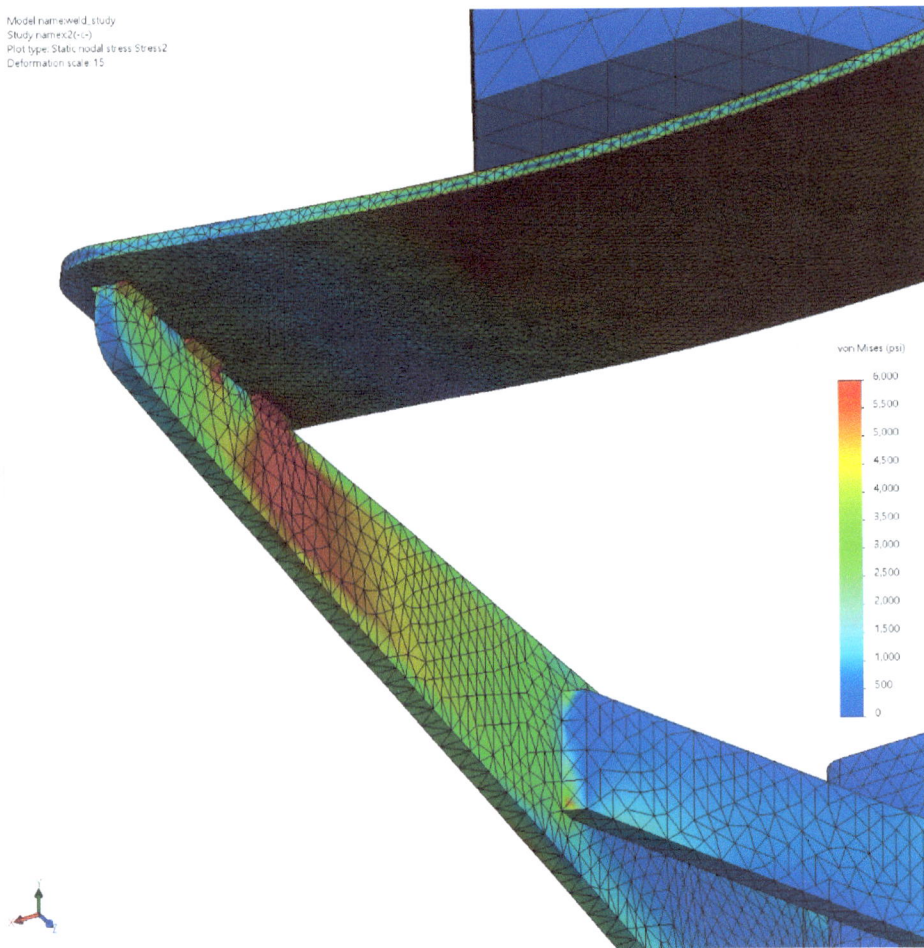

Result with no penetration contact (6 ksi scale)

Note that the mesh in the table had been refined to give two elements across for accurate bending response. More on this in a later paragraph.

We mentioned at the start several advantages of modeling the load. The payoff follows. The gravity vector is put at an angle, to emulate a lateral acceleration. With a 30 degree shift in angle, 0.5 g lateral acceleration is expected. Friction in the no penetration contact with the load is increased to make sure it stays put.

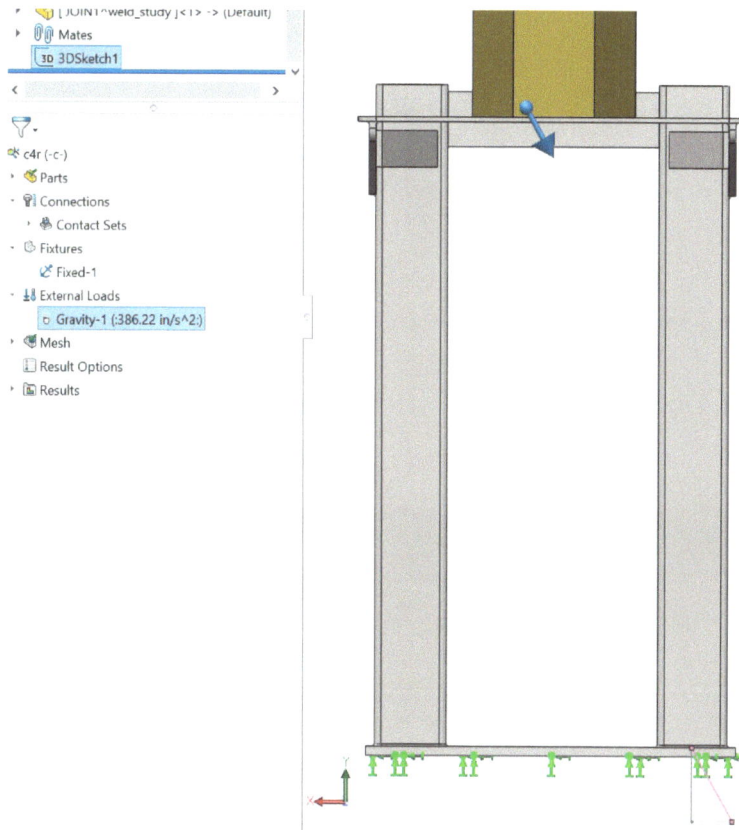

Study setup with angled gravity vector, along sketch line

Surface Mesh Refinement

Significant bending and twist is expected in the arms, so a mesh refinement is made on their flat surfaces at one half their thickness.

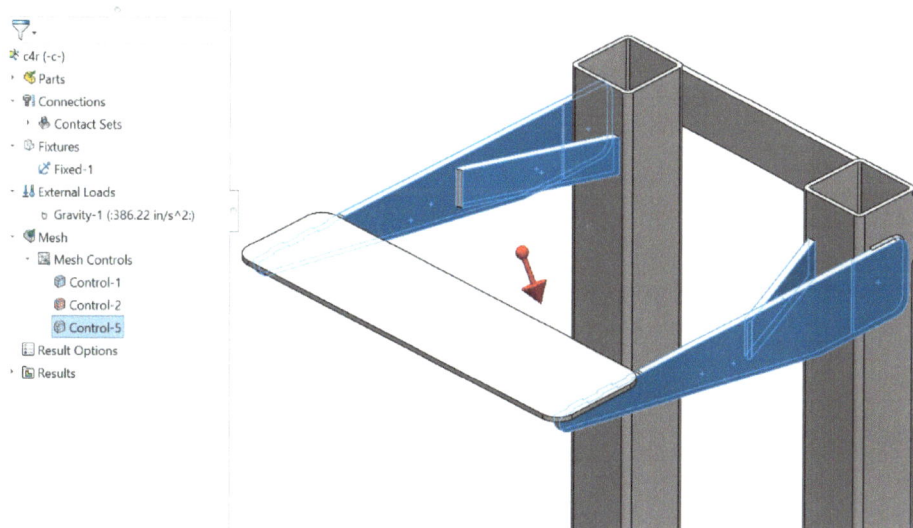

Surface mesh refinement

We have had good results refining opposing surfaces of flat plates. The surface refinement leads to a more regular mesh than a refinement applied to the entire volume. Mesh sectioning will confirm this visually. Surface refinement also makes fewer elements for the same general density.

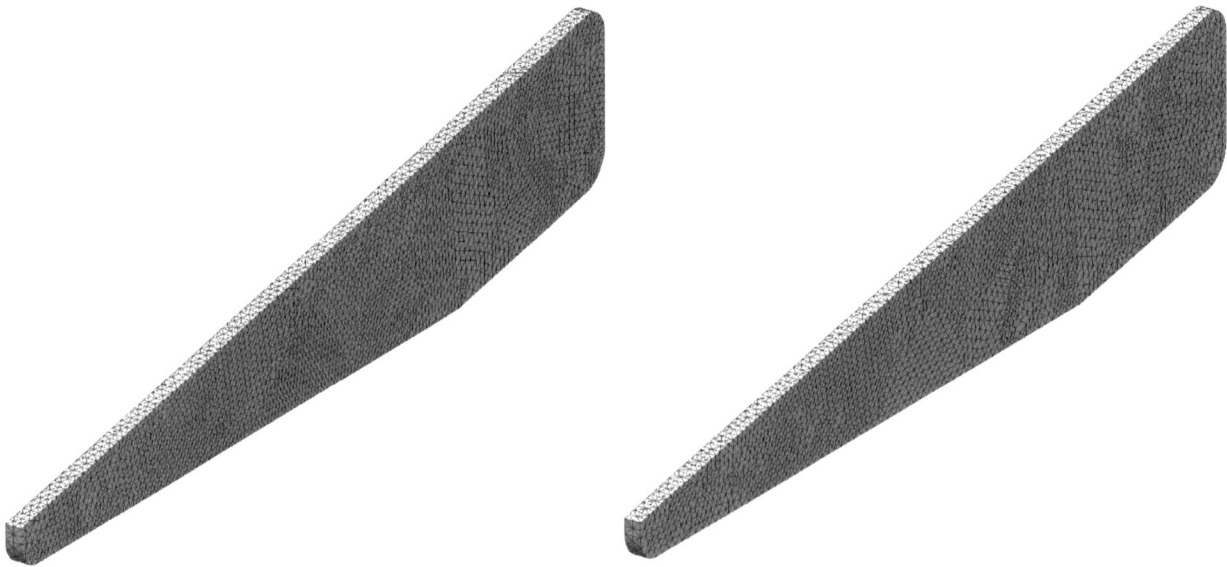

Mesh on separate arms, volume refined vs surface refined

The differences here are not blindingly obvious. But the volume refined model has 41508 elements versus 31429 in the surface refined part. In a simple bending load case they deflect to within .0027% of each other.

The curious user can easily set up a series of studies to see the benefit and cost of using one, two, four, or even eight elements across a plate thickness, first-order (draft) or second-order (Solidworks default) or even higher order. When the plate is loaded in bending and twist, two second-order tetrahedral elements through the thickness is good for most purposes we have encountered.

Now with the mesh refinements and high accuracy no-penetration contact study run time is up over 11 minutes on a modern laptop computer. For that price we have a model with every weld detailed and high confidence in the result.

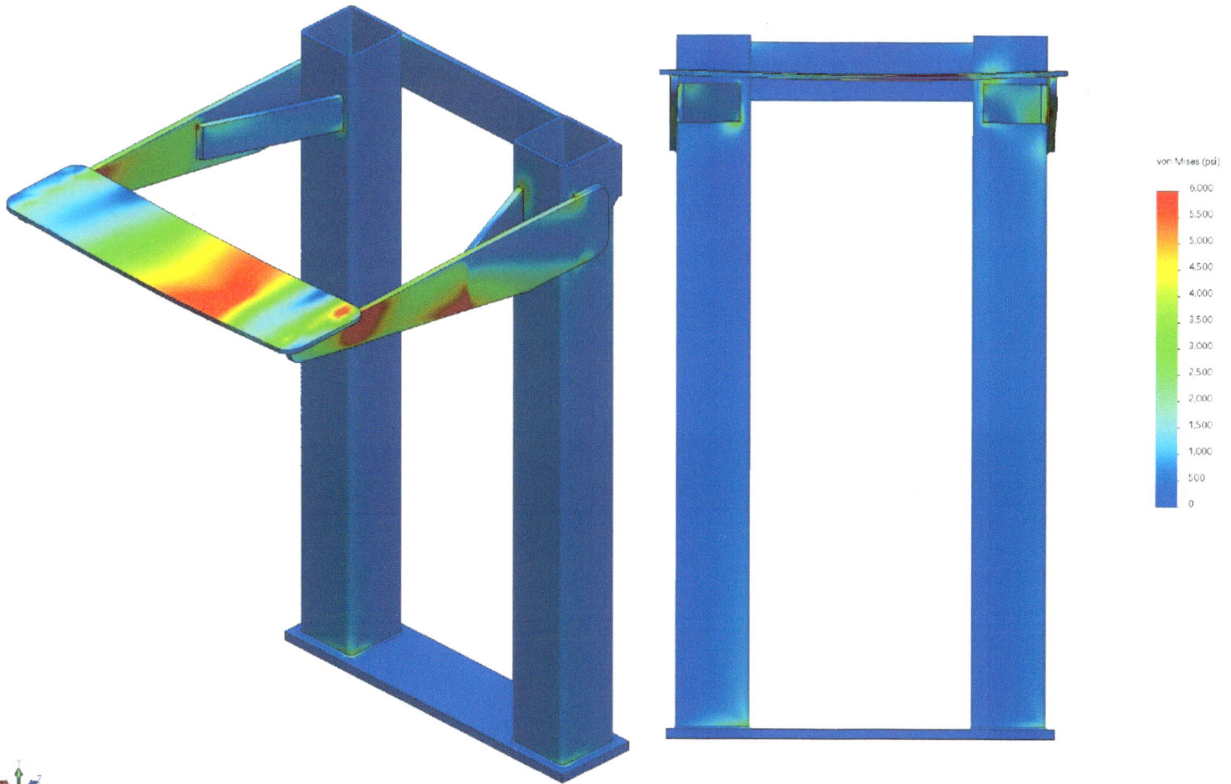

Model name:weld_study
Study name:c4r(-c-)
Plot type: Static nodal stress Stress1
Deformation scale: 5

Results with lateral acceleration (6 ksi scale, 5x displacement magnification)

Of course we have been using a severe color scale and displacement magnification. We can put the scale at 2/3 of the lowest material yield point and document this machine as ready to build.

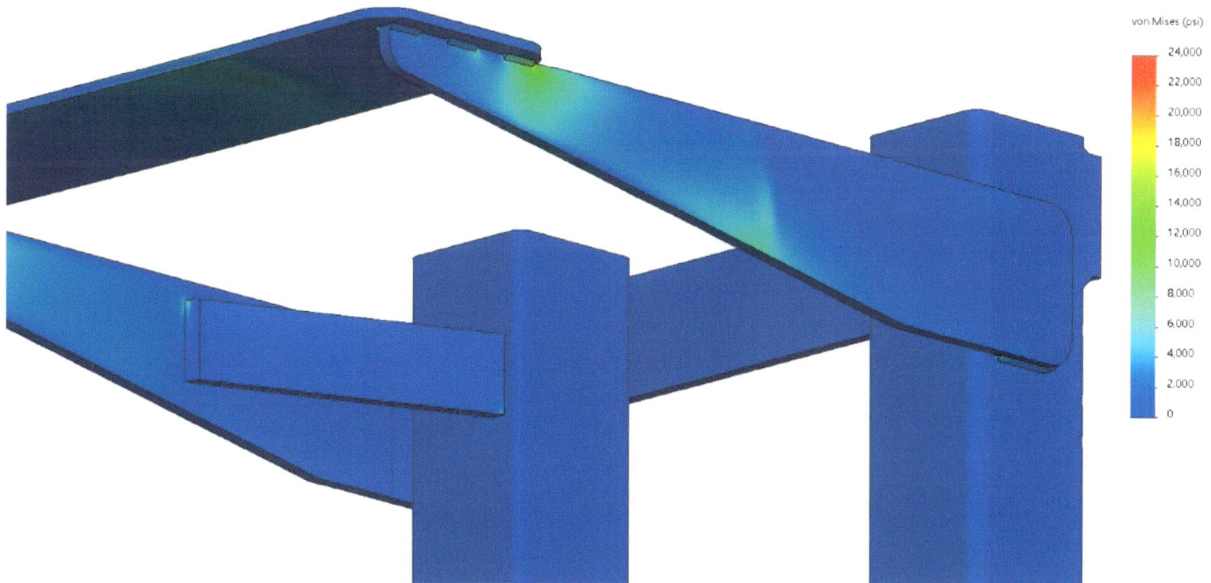

Result with normal stress scale, true displacement

Conclusion

We hope that in this guide the engineer or analyst has found tools helpful in design of successful weldments. Much finer detail around weldment joints is needed than what is seen in typical software demo models. The techniques shown here have proven themselves to allow relatively efficient use of labor and computer resources, while providing the design insight needed.

Contact Us

Contact Stone Lake Analytics with any questions or comments. We are a full-service design and analysis consultancy with specialties in casting dies, articulating machines, and welded structures. www.stonelakeanalytics.com.

Author's own welds, 1" plate brackets for supporting 55,000 lb. test fixture.
(Author is not an elite welder, but has not had anything fall on him yet)

Solidworks Simulation for Real Weldments

www.ingramcontent.com/pod-product-compliance
Lightning Source LLC
Chambersburg PA
CBHW052048190326
41521CB00002BA/146